The Technology War

A Case for Competitiveness

David H. Brandin

Strategic Technologies, Incorporated
Los Altos Hills, California

Michael A. Harrison

University of California
Berkeley, California

A WILEY-INTERSCIENCE PUBLICATION

JOHN WILEY & SONS

New York / Chichester / Brisbane / Toronto / Singapore

Ada is a registered trademark of the United States Department of Defense, Ada Joint
    Program Office.
IBM is a registered trademark of the International Business Machines Corporation.
Sun Microsystems is a registered trademark of Sun Microsystems Inc.
UNIX is a registered trademark of Bell Laboratories.
Quotation from *Iacocca: An Autobiography* by Lee Iacocca with William Novak copyright
    © 1984 by Lee Iacocca. Published by Bantam Books, Inc. All rights reserved.
Quotation from *Hitler's Spies* by David Kahn copyright © 1978 by David Kahn. Reprinted
    with permission of the publisher.
Quotation from "Hicks Attacks SDI Critics" by R. J. Smith copyright © 1986 by the AAAS.
Quotation from "In Europe, the 'Bunch' Hope to Connect to Top Banana IBM" by Philip
    Hunter copyright © 1986 CMP Publications, Inc., 600 Community Drive, Manhasset,
    N.Y. 11030. Reprinted with permission from *InformationWEEK*.

Copyright © 1987 by John Wiley & Sons, Inc.

All rights reserved. Published simultaneously in Canada.

Reproduction or translation of any part of this work
beyond that permitted by Section 107 or 108 of the
1976 United States Copyright Act without the permission
of the copyright owner is unlawful. Requests for
permission or further information should be addressed to
the Permissions Department, John Wiley & Sons, Inc.

Library of Congress Cataloging-in-Publication Data

Brandin, David H.
  The technology war.

  Bibliography: p.
  1. High technology. 2. Computer engineering.
3. Competition, International. I. Harrison, Michael A.
II. Title

T49.5.B736   1987   338.9'26   87-6127
ISBN 0-471-83455-6

Printed in the United States of America

10 9 8 7 6 5 4 3 2 1

To

Tohru Moto-Oka

(1930 – 1985)

Pathfinder of the Information Age

Distinguished Computer Scientist and Colleague

Preface

The Technology War is about the worldwide race to capture the lead in *the* strategic technology: *information technology*.

The horror of nuclear war has rendered thermonuclear conflict unthinkable. It does not and cannot serve as a rational basis for satisfying national strategic objectives. Conventional war might still meet limited tactical goals and can resolve conflicts involving third-world or nonnuclear states. But because conventional war among superpowers is likely to lead to nuclear escalation, it must ultimately be abandoned.

Technology, because it affects standards of living and wealth and because it is the source of the superweapons of the future, is the new arena of competition. But this is not the traditional competition of capitalism. This conflict is more sophisticated than the competition found in most of the existing political and economic systems. Only the most clever and enduring competitors will be rewarded in this arena. Those that prevail in this war will control the resources of the world; they will control their *Lebensraum*; they will be the next global powers.

Forty years after winning World War II, the United States is still the leading world power. And we should be able to expect this to remain so. After all, the United States is the leading producer of agricultural goods, it enjoys a large and prosperous middle class, and it is rich in natural resources. Given the facts that it has the most powerful economy in the world, has a brisk and vigorous technological community, and has an industrial base that was unmauled in the Second World War, the United States should be able to maintain its dominance indefinitely.

That leadership is at risk. Asian countries have experienced growth rates three times that of the United States. The Vietnam War, which almost bankrupted the country, also demoralized it. The artificial prosperity of the "guns and butter" economy made the middle class complacent. Behind the smugness of U.S. supremacy, the country lost its monopoly on technology. The result is unacceptable levels of unemployment and the loss of jobs – middle class jobs – to our competitors. These jobs are one of the few common denominators in measuring the state of the Technology War.

Unfortunately, this war cannot be analyzed in the context of traditional wars. There are no simple measures for counting weapons; no easy methods for determining the presence or strength of enemy forces. It is not even clear who the real enemy is! Yet the competition in information technology has already altered the balance of power in the world.

Information technology is a sophisticated field that goes far beyond the science and engineering of data processing. Managing its development requires an informed leadership that recognizes that even seemingly benign issues are vital elements to be coordinated in the nation's strategic arsenal. The relationships between industrial policy, the intellectual property laws, capital and tax policies, government research programs, trade policies, and technological development are complex. Access to and the exploitation of the leading edge of technology are also important.

Having access to the technology, through its development, however, does not necessarily lead to success in exploiting it. In this, education and other social forces are also determining factors. Accordingly, this book about technology is also a book about economics, education, sociology, and international competition. To understand the significance and the value of the high stakes in this race, we must understand all of these relationships. The Japanese and European governments, and a host of other nations and organizations have already perceived the need for a competitive edge in information technology. They believe it is the key to the education of their children, the competitiveness of their industries, the well-being of their aging populations, and the security of their nations. It is a competition they take seriously.

The competition is manifested in a variety of major and minor research and advanced development programs, such as the United States Strategic Computing Initiative (SCI), the United States Strategic Defense Initiative, the Japanese Fifth Generation Computer Technology Program, and the European Strategic Program in Information Technology (ESPRIT).

Within the United States, several studies of these competing programs have been completed. We have participated in and directed two of these studies. David H. Brandin was chairman of the Department of Commerce's Japanese Technology Evaluation Program on Computer Science. Michael A. Harrison was chairman of the National Research Council's Panel on International Developments in Microelectronics and Computer Science. These projects and other studies have enabled us to study information technology in Japan, China, and Europe, as well as in the United States. Many of the concepts developed in this book are consequences and extensions of those studies.

In our investigations and in the preparation of this book, we had the assistance of many people and organizations. We specifically want to thank Ian Buchanan, Pehong Chen, John Coker, Adele J. Goldberg, Jean-Paul Jacob, Michael A. Jacobs, Robert E. Kahn, Erin A. Kelly, Joe Kelly, Hishashi Kobayashi, Franklin F. Kuo, Jeff McCarrell, William F. Miller, Clyde Prestowitz, Steven Procter, Olivier Roubine, Harry Rubin, George Telecki, and Fran Peche Turner, the Association for Computing Machinery, Cezar Industries, Ltd., the Computer Science Division of the University of California at Berkeley, the National Academy of Sciences Panel on International Developments in Microelectronics and Computer Science, the National Science Foundation (Grant MCS-8311787), the U.S. Department of Commerce Japanese Technology Evaluation Panel, the U.S. Department of Defense Advanced Research Projects Agency (DARPA Order No. 4871, monitored by Naval Electronic Systems Command, Contract No. N00039-84-C-0089), SRI International, and, most especially, our wives, Ellen Brandin and Susan Graham Harrison, for their enthusiastic support of this project.

David H. Brandin
Michael A. Harrison

San Francisco, California
June 6, 1987

Contents

PART I
THE TECHNOLOGY WAR

1

The Technology Imperative

The chairman of a powerful British company publicly attacks the United States for its repressive controls of high technology information. Japan declares its intention to "lead IBM, rather than follow." Despite unquestioned American supremacy in artificial intelligence, the Japanese doggedly announce their intention to found their new computer technology on artificial intelligence and knowledge-based engineering.

The United States responds to the Japanese announcement with an aggressive research program managed by the Department of Defense. The U.S. Department of Justice blesses the creation of industrial research consortia that ten years ago would have provoked the fury of the antitrust laws. The U.S. secretary of commerce proposes to abolish a provision of the antitrust law. The European Economic Community (EEC) establishes ESPRIT, the European Strategic Program for Research in Information Technology. ESPRIT meeting attendees accuse the United States of exploiting the International Traffic in Arms Regulations to deprive the Europeans of information technology products and engineering. Such deprivation is, presumably, to protect the market positions of American suppliers of information technology prod-

3

ucts. Britain establishes a program in Advanced Information Processing and restricts participation to residents of the United Kingdom.

What has happened to the traditional concepts of openness and cooperation in science and engineering? Are they casualties of an undeclared war? Can we expect to see a continued tightening of the export controls on high technology, the closing of our nation's doors to foreign scientists?

The answers to these questions are yes. They are the early signs of a scientific and economic competition that has replaced the warfare of earlier years. This battle over competitiveness involves money, national wealth, pride, the standard of living, and the future well being of the citizens of the competing countries. It involves the future of the world.

Clyde Prestowitz, a former Department of Commerce official, has called the competition in semiconductors between Japan and the rest of the world the "Chip Wars." A war, he wryly notes, that is already lost. The EEC, in their recent imposition of an unprecedented tariff of 16 percent on photocopiers, referred to the Japanese dumping of copiers in Europe as the "Photocopier War." Rob Wilmot, the former Chairman of ICL, a British computer firm, was no less direct when he referred to the "fierceness of battle between the U.S. and Japan with Europe caught in the crossfire." Indeed, even the Eastern Bloc perceives a threat through its usual bureaucratic stupor: "The Socialist Community of Nations," they say, "is now facing the greatest challenge in its postwar history: The Technological Challenge." This is not just the early skirmishes in a trade war, this *is* a war over technology. It's a war the United States may be losing.

Whether we fight or not, the consequences of losing the war are the loss of national wealth, prosperity, leadership, employment, national security, and freedom. It means more smokestack industries will decay. For the United States, it means no less than what the British experienced after the Second World War: increasing unemployment and unrest, a debasement of the currency, and the humiliation and the shame that accompany the loss of power. The successful outcome to this war is critical to the well-being of America.

In the Technology War, we need to reconsider who our principal adversaries are. Indeed, our trading partners and our allies, Japan for example, are our fiercest competitors. Curiously, our principal adversary since World War II, the Soviet Union, is *not* a serious adversary in the Technology War. There is very little flexibility in the Soviet system to deal with its basic weakness in information technology. Comparatively speaking, there also may not be enough flexibility in the Western systems of governments to exploit the leading edge of technology effectively. But the Soviet bureaucracy, widely known for its clumsiness and ineptness, is a far deadlier anchor, and it will surely sink the Soviets, in the Technology War. We believe it is too late for them to catch up with the United States and Japan.

In an attempt to keep pace with the West, the Soviet Union has decided to introduce computer technology into its educational system. This is quite a concession since a knowledgeable and computer literate citizenry poses a threat to the traditional Soviet centralized control of information. In teaching computing, the Soviets now are acknowledging that the West has pulled far ahead of them in computer science.

Actually, the poor performance of all Soviet scientists is bewildering. The 1.5 million Soviet scientists constitute one-fourth of all the world's scientists. Yet, they have won only 11 of the Nobel Prizes awarded in science this century as opposed to the United States' 145 prizes. Furthermore, the Soviets have won only two of the science prizes in the last ten years. They have only one-tenth the number of mainframe computers as in the United States. In contrast to other fields where they publish extensively, such as mathematics, their lack of publications in information technology supports the argument that they lag the West by a good generation of technology.

The Soviets' emphasis on military technology and defense priorities has diluted their effort on other applications of information technology. For example, eliminating the Soviet direct distance dialing network in 1982 to facilitate KGB monitoring of telephone calls seriously impaired the Soviets' development of modern digital communications, a necessary ingredient in the commercial development of information technology.

Economic Power

The economic importance of information technology is astonishing. Indeed, the information technology annual marketplace in the late 1980s has been conservatively estimated to be $100 billion. The software market alone is expected to reach $60 billion by 1988. The information technology marketplace is also a major producer of jobs, and total employment in the field is estimated at 1.5 million in the United States and 2.5 million worldwide; estimates that we believe are understated.[1]

Moreover, information technology has created jobs faster than any other field; a fact grasped both by our major trading partners and by the Third World countries. The Europeans have noted, for example, that if their technology markets had developed at the same rate as those of Japan or the United States, about 2 million more jobs would have been created in a community with unemployment at about 12 million in 1985.

It should be clear that the real prize in the technology race is economic power. But, the economic implications are somewhat difficult to understand because each economic and social system has a different set of incentives. Furthermore, the definitions of victory are different. The American desire for quarterly bottom-line earnings contrasts with the Japanese willingness to trade profits for marketshare and employment. Comparisons between competitors that are based on investment levels are shaky because of the different savings rates, tax laws and incentives, regulatory forces, and so forth. Still, the prize of economic power is measured best by marketshare and to a significantly lesser degree by profits.

Defining the information technology marketplace is no small task. Experts disagree vehemently on its size and structure.

Should cameras be included since new electronic imaging technology exploits microprocessors and floppy disks? Do we consider smart microwave ovens part of the market? Should automobiles that talk and computerized navigational systems be counted?

Further, communications technology is heavily dependent on computers, and the interface between computing and communications in many systems has become blurred. The number of people selling services related to this field has expanded beyond our ability to measure it. As a result, the role of communications and services is ambiguous and has an influence of perhaps 100 percent on the potential size of the market. The U.S. Department of Justice hinged its most recent antitrust actions against IBM on this issue and lost, in part because the government was unable to develop a convincing definition of the marketplace.[2]

Because the profits that finance research derive ultimately from marketshare, the interdependence of marketshare and basic research is important. We cannot be complacent over a slow loss in marketshare because the accumulating effect on research and development, that is, the shrinking availability of profits to finance new research, will ultimately accelerate the loss. Because of the amount of time it takes for this feedback effect to be realized, major national strategic objectives will be compromised or sacrificed without any direct warning. Thus, there is cause for real concern over our slow but steady loss of high technology marketshare to the Japanese and Koreans.

In fact, those with access to and control of information will be the power brokers of the future. They will become increasingly powerful and information wealthy at the expense of their competitors, creating an information technology gap. It has been suggested that information wealth is a new type of capital known as *knowledge capital* and that it may be more important in the future of the U.S. economy than ordinary capital. Economists have not paid much attention to this new type of capital when evaluating economic performance.[3] Perhaps they should. The combination of information technology-based industries and services should generate a $1 trillion information processing industry by

1995. Can any other industry lay claim to a greater potential impact on the nation's future?

These commercial implications of information technology are terribly important to the well-being of the United States. They impact on the wealth and the economic security of the people. However, from some perspectives, they pale in comparison to the effect that information technology has on national security.

National Security

Although the computer was useful for artillery calculations and other numerical applications in World War II, at that time almost no one recognized its enormous potential for ensuring national security or the role the computer would play in America's modern defense forces. The failure to grasp the value of technological achievements is not unusual. In fact, in 1936 few people sensed the potential for the rocket or believed it would play an important role in warfare.[4] Most pundits predicted, for example, that most of the data processing requirements of the country would be satisfied by four or five machines of the UNIVAC I class (a first generation computer whose capacity is exceeded by today's sophisticated hand-held systems). These early computers were useful for calculating nuclear weapon design characteristics, lunar trajectories, and so forth.

In 1978, President Carter's Federal Automatic Data Processing Reorganization Study reported that the Department of Defense is incapable of performing its mission without computers, specifically that we cannot fly an airplane, maneuver a tank, or fire a modern weapon without the assistance of computers. Since then, the use of computers in defense has become pervasive. The ubiquitous computer is used in the control of forces, targeting, intelligence, and even automotive systems. Yet despite this enormous dependence on computing, there is skepticism about whether the U.S. Army is capable of rapid mobilization because of inadequate data processing capability to handle the expanded payroll requirement that would be created.

In today's world of cold war confrontation, covert operations based on intelligence data have become, perhaps, as important to the nation as the use of conventional forces. In that arena a continuous supply of state of the art computers has helped the military meet its objectives. Since World War II, computers have served as the critical encryption and decryption tool. They have become the only means to deal with the enormous quantities of classified and sensitive communications data that must be protected or intercepted and decoded. The National Security Agency, one of the preeminent users of computer technology, would have to abandon its job if its computer facilities, perhaps the largest in existence, were unavailable.

Up to 25 percent of the defense budget is intimately related to the acquisition, management, or use of information technology. Each new sophisticated weapon system carries a critical component of information technology. The new Strategic Defense Initiative, for example, requires the use of computers to track, detect, discriminate, manage fire control, and steer, in real time and without fault, all of the myriad countermeasures. Some of the components are presumed to be completely under the control of automated systems.

The North American Aerospace Defense Command is totally dependent upon the use of computers to track satellites in space and monitor space-based threats to America's strategic forces. Absent this computing base, forces such as the Strategic Air Command would be defenseless.

Other government agencies also depend critically on the use of computers and improvements in the technology. Applications such as air traffic control, earthquake prediction, and weather prediction can all benefit from improvements in microelectronics, supercomputers, and software technology. The air traffic control computer system, for example, is aging, causing problems in reliability, capacity, and safety. The Federal Aviation Agency will spend hundreds of millions of dollars to replace this fleet of computers with the latest commercially available equipment. These latest computers, however, will be based largely on the research of the 1960s and 1970s, such as networking, fault tolerance, microelectronics, architecture, and software engineering. And because applications such as those mentioned will always need more computing capacity, the research of the 1980s must provide for the 1990s and beyond. The way in which we apply the research results will have enormous impact on our safety, future leisure, wealth, job opportunities, and national security.

The Technology War

The world's greatest high stakes science race has started. This race is contagious and has spread rapidly to the corners of the world. Each nation-state is engaged in the battle—some as combatants, others as spectators. The great industrial powers are massing large resources in their research laboratories. Even cities vie for advantage against each other. Austin, Texas, for example, competed vigorously with San Diego, California, and Winston-Salem, North Carolina, for the privilege of serving as the home of a flagship research consortium known as the Microelectronics and Computer Technology Corporation. Some research scientists sell their services to the highest bidder, much like mercenaries, driving up the cost of research and raising the stakes. Terrorists have murdered the managers of research, spreading the strife.

There have been earlier competitions. From the days of Aristotle to Hitler, scientists have waged war on each other. The First World War saw the use of poison gas, tanks, and submarines. During World War II, Allied and Axis scientists sought to outsmart each other in such diverse scientific areas as cryptography, radar, materials, navigation, and rocketry.

Societies have also waged war on science. Socrates was forced to drink hemlock for filling the minds of the youth with philosophical thought, and Galileo was persecuted by the Church for his astronomical heresy. In this century, the Chinese Cultural Revolution deprived an entire generation of intellectual and scientific development.

In this book we describe the current Technology War and

provide insights into its status and likely outcomes. In the next chapter, we introduce the major combatants and their important strengths and weaknesses. This is exemplified by the erosion of America's marketshare as the Japanese exploit their advantage in applied engineering.

The battlefield in this war has several fronts. The first is the technology manifested in hardware, software, and the exciting new fields of microelectronics and artificial intelligence. A description of the important components of these technologies is included followed by an introduction to the other strange terrain in the conflict: technology transfer, government and industrial policies, and education. Herein lies an examination of the American and European ineptness at managing their resources and the admirable Japanese ability to exploit their, as well as our, systems.

America's position as the leader of the Western alliance is clearly at stake in this conflict. Our leadership in science and technology is especially precarious. This book concludes with some suggestions for a strategy for the United States, one that might reverse its present course toward second-class status in the world.

2

Winners and Losers

Many factors, including societal, educational, and financial ones introduce great uncertainty into determining who is winning the Technology War. Because America, the nations of Europe, and Japan all have different value systems, it is difficult to estimate one nation's progress or to compare the competitive positions of several nations. Each country places a different value on its different objectives, often making it necessary to compare apples to oranges. To the Japanese, for example, employment is more important than profit, whereas to the Americans, profit is the most important.

Another factor that prevents us from measuring progress in the Technology War is that new technology reports involving developments in many of the latest commercially relevant fields are no longer being published in the open literature. Each of the competitors has moved toward greater secrecy in order to protect its ideas and innovations. This is a phenomenon with which most scientists are unfamiliar. Physics, for example, has been relatively easy to track through the last few centuries simply by reading the technical papers. But the new technologies, such as biotechnology and, especially, computer science, are less mature fields and access to their publications is not adequate to explain the state of the art.

In addition, most high technology companies reinvest 5–10 percent of their sales in research. This creates research budgets ranging from $500 million to several billion dollars for industry giants such as IBM and NEC. Although, a large percentage of the research budgets (often as high as 90 percent) is devoted to advanced development (sometimes a euphemism for product engineering), much of it is for basic research. In fact, these proprietary research budgets often overshadow government-funded programs. We must analyze an entire national effort, including government, academic, industrial, and private projects to obtain a meaningful comparison of the different national programs.

Similarly, the markets affected the most by the information technologies are changing rapidly. Some product life cycles have shrunk from the traditional engineering experience of eight to ten years to less than one year. Some technologies have been totally replaced by newer ones, such as vacuum tubes by transistors. It is difficult, indeed, to state the relative positions of competitors. One attempt to establish a comparison was made by the United States Department of Commerce.

Technological Comparison Between the United States and Japan

In 1984, the Department of Commerce organized a Japanese Technology Evaluation Program (JTECH) in which they set up several panels of technical experts. These panels were able to compare the two nations in four technical specialties and in three general areas: basic research, advanced development, and production engineering.

The four studies were Computer Science, Micro- and Optoelectronics, Mechatronics (the Japanese label for Robotics), and Biotechnology. With few exceptions, the findings of the four studies were similar; they are discussed below.

After a year of visits, literature searches, and working meetings, the panels reported that there was a definite pattern across the spectrum from basic research to advanced development to product engineering. In basic research, it was found that Japan was far behind the United States and slipping further behind. In advanced development, the U.S. advantage was not quite so strong; Japan was considered behind the United States but holding its relative position. The U.S. advantage disappeared, however, in product engineering, where the Japanese were considered at parity with the United States. Most importantly, the Japanese were improving their position in manufacturing at the expense of the United States.

We believe that in the late 1980s the Japanese are ahead in product engineering, especially in manufacturing. In device development in optoelectronics, the Japanese are also superior to the United States (although the United States still leads in some optics systems areas because of its substantial lead in the critical field of software). In another new field, biotechnology, the Japanese research activity is on a par with that of the United States due to substantial government participation that began in 1981. In robotics, Japanese basic research is considered equal to that in the United States, with the exception of computer vision and software, which are widely acknowledged to be bastions of American strength.

In all the categories, the Japanese have an increasing advantage as we move from research into production. This shift in the relative advantages is explained by the differences in industrial policy and research investments in the two countries. The United States has traditionally placed its government emphasis on research, whereas Japan has regularly concentrated on advanced development. Thus, in the last forty years, the Japanese have established a clear trend toward superiority in manufacturing.

Japan's benefits have included rapid economic growth and market penetration along with high employment, but at the expense of profits. The Americans, on the other hand, who were dominant in manufacturing earlier this century have since shifted their emphasis on innovation from manufacturing to research, at the expense of world markets. This was not a conscious decision, but rather a case of falling asleep at the wheel.

In the twentieth century, European manufacturers have not been willing to sacrifice profits to sustain employment. Rather, they have acceded to their shareholders' demand for dividends. There has been no incentive to invest for the purposes of expanding their capacity to meet future demand. They failed to modernize their industrial plants (only Germany, after the obliteration of its industrial plants in World War II, enjoys the benefits of a modern industry in Western Europe today). The result is a dismaying loss of competitiveness for European manufacturers that can only be compensated for, as Barzini said, by "[local] bureaucracies that continue to invent ever-ingenious ways to stop the importation of some goods from other ... countries."[5] Unfortunately, U.S. stockholders have been equally resistant to the diversion of profits to reinvestment. The resultant shift already can be illuminated by a comparison of factory production indexes. Using 1957-1959 as a base of 100, in just ten years, Japan has improved to 402 whereas the United States had a modest increase to 172 and Great Britain crept up to 137. These indexes are proportional to the improvements in the standards of living in each nation.[6]

In this multifaceted war, each competitor has advantages and disadvantages. For example, the Japanese have been more successful than the West in exploiting their favorable position in the creation of capital. Reischauer has pointed out this critical relationship between trade, technology, and capital:

> [A] great flow from the United States to Japan of industrial technology ... and substantial amounts of American banking capital ... contributed critically to Japan's economic recovery and the restoration of its world trade.[7]

This recovery is demonstrated painfully in the microelectronics markets. American industries were once the dominant force in the development of microelectronics and random access memories (RAMs). U.S. suppliers essentially controlled the entire market.

But during a medium-scale recession in the United States, American vendors, preoccupied with bottom-line earnings, retrenched on their capital investment program in new facilities. The Japanese, on the other hand, with an eye toward future markets, made great investments in capacity during that time. As a result, Japan was able to meet the postrecession demand. Their share of the market grew dramatically at the expense of American suppliers. American semiconductor manufacturers now plead for a "level playing field" in their competition with the Japanese. Indeed, Japan is now the dominant force in RAMs.

In contrast, the United States is expected to dominate basic computer science and most other fields of research through the next decade mostly because of its present base and momentum. The Japanese are the only serious commercial competitors to the United States but they have been almost totally dependent on basic research done outside Japan. To compensate for this weakness, the Japanese are clearly moving into a research phase in these critical technologies. In fact, this research phase was initiated about 1970 but did not become noticed by the West until a decade later.[8] A tendency to discount the potential for research in Japan is probably what led to the ten-year gap in detecting the start of Japanese interest in research in the West. Certainly, the language barrier contributed as well. In addition, it was not in the Japanese interest to proclaim this new focus on research. Their low key strategy worked to their advantage under the umbrella of Western arrogance. Still, it was not a well-kept secret. The Japanese have been very much in evidence at the major international scientific conferences, and their presence in technical programs has been increasing regularly. The United States has no one to blame but itself for ignoring the evidence.

The United States

Still, the United States is the dominant force in the overall Technology War. After all, the United States represents the largest domestic market, has had the greatest percentage of its GNP devoted to research and development, and has enjoyed the fruits of a massive defense research program since the 1940s. In fact, the computer was developed with defense funds in World War II to meet the needs of the War Department and the British Ministry of Defense. Cryptology, artillery, and the atom bomb projects generated extraordinary demands for computational services that continue today to command the government's interest in computing. Further, the United States has the largest number of independent firms participating in a vigorous, competitive and innovative climate. Such competition has resulted in tremendous technological accomplishment.

Computer Science in the United States

Indeed, the United States has long been and continues to be the world leader in research into computer science. The computer science field has benefited consistently from our continuous program of investment. In spite of the much heralded Japanese Fifth Generation Computer Program and the other Japanese research efforts devoted to computer science, the United States has far larger commitments and a substantial and healthy research tradition that has led to a great American advantage in this field. Further, as a result of over 40 years of experimentation with information technology, the United States has developed a substantial advantage in software technology.

This advantage is under attack by our competitors because, as the relationship between hardware and software costs has changed, software has become much more critical. With the declining hardware costs, software costs now dominate total system costs in most applications of computing, whether they are for off-the-shelf systems or for newly developed or engineered systems.

There is no simple formula to determine the costs in the development of software. Software is labor intensive, and software costs do fluctuate in different labor markets. Such fluctuations might create a variety of opportunities for our competitors. De-

veloping countries with lower labor costs, such as the Republic
of Korea, Singapore, Taiwan, India, and Malaysia, may, through
their fine educational systems and knowledgeable work forces,
find a competitive advantage in some software markets. They
have already targeted software as a critical technology and market opportunity. Also, Japan, through an advantage created
by the combination of cultural and work force practices, seems
to generate more reliable application software than that of the
United States. With this advantage, they might find a way to
challenge the American lead in some software markets. For example, the Japanese Technology Evaluation Program reported
that the Japanese can develop production software that is up to
ten times more reliable than American software.

However, the special nature of the Japanese software market
mitigates against their apparent advantage in developing more
reliable software. Japanese users are not accustomed to using
general-purpose software; they demand applications tailored to
each customer. This creates a tremendous demand on the Japanese resources available for programming. McClellan has written that "Software may prove to be the greatest barrier to the
Japanese ... because Japanese customers have preferred custom
software for a long time, there is no large body of standard software packages."[9] On the other hand, the United States does not
suffer as acutely from this programmer productivity problem in
the bulk commercial software marketplace. As a result, a larger
percentage of American programmers are available for creative
work and the exploratory development of new software concepts.

In evaluating any form of productivity, we must also consider
the value of the currency. For example, we have been led to believe that Britain has been systematically outperformed by the
United States. Indeed, after World War II, Great Britain's economy deteriorated and it found itself mired in a permanent state
of 12 percent unemployment. But despite this problem, Great
Britain was able to compensate for the poor performance of its
economy and the sharp increase in its labor costs by substantially devaluing its currency relative to the United States. As

a result, the United States was actually the worst performer in labor productivity in that period.[10]

Japan

Japan is the most aggressive competitor confronting the United States in the Technology War. With over 180,000 installed computer mainframes (worth about $37 billion at today's prices), it has the second largest domestic economy in the world. However, Japan's appearance in the 1970s as a strategic competitor in information technology was a surprise to most American observers. They were accustomed to thinking of the Japanese in terms of their experiences in the 1950s and 1960s. After all, Japan had been devastated in World War II and was considered to be a desperate low cost producer of shoddy goods. No one was prepared for the technology shock of the fifth generation computer announcements.

The first moments of awareness in the United States and other Western nations came in 1981. Japan hosted an International Conference on Fifth Generation Computer Systems. The keynote speaker, Professor Tohru Moto-Oka, described his ambitious vision of society and the Information Age in the 1990s. Since the state of information technology was and is well below that level, a major research and development program was being initiated by Japan. Moto-Oka suggested that Japan would lead the world toward the golden age of the Information Society by the exploitation and continued development of computer science and artificial intelligence. Moto-Oka's vision resulted in the first and most original of the worldwide research programs. In a sense, this was the opening volley in the Technology War.

This research program is only one of the more well known Japanese activities among a number of important advanced development areas in the commercial sector. Since Japan does not have the burden of large defense expenditures and the direct social obligations of the United States, they are able to focus their research on their most pressing domestic and commercial needs.

This ability to focus their efforts is a part of the Japanese comparative advantage in technology transfer and is dependent on the characteristics of their unique society.

Characteristics of the Japanese System

One of Japan's strengths in the Technology War with the West is its encouragement of a highly competitive yet cooperative society. Commercial success is important but not at the expense of society. Japanese businessmen are trained to cooperate, in a sense, through their studies of military principles. The works of two oriental authorities on military strategy, Musashi and Sun Tzu, for example, are required reading, at every Japanese secondary school.[11] These teachings place great stress on the understanding and use of all available weapons needed for success, as well as on recognizing the need for cooperation.[12]

The Japanese socioeconomic system, in a process continually refined by the government, encourages all interested parties to develop a common competitive base. This can be seen in the large number of advanced technology development programs sponsored by the Ministry of International Trade and Industry (MITI), in which the participating companies share all the research and development results in what is known as a precompetitive phase. This phrase is finding increasing use in the United States and Europe where corporations have found it useful to adopt similar practises by creating cooperative research programs.

The practice of cooperation is true of individual researchers as well. Few Japanese scientists desire to be known as individualists or to stand out from the crowd. It is common for the scientific community in Japan first to build a base of known ideas and then to exploit those ideas developed from other researchers. They say, for example, "through the study of Western works, we intend to leap-frog the predecessors."[13] The Japanese do not view the approach of imitation and subsequent innovation as being contradictory; indeed they feel "they are handled as a

series of ideas: copying the idea will eventually lead to original-
ity as the next step."[13] Rarely do the Japanese deviate from this
consensus, research, imitation, and innovation sequence.

We often hear the argument that the Japanese lack creativity and must ultimately suffer in competition with the West. We
believe, however, that the Japanese can perform world class research; which until now they simply have not had a need to focus
on. JTECH, after studying the issue for over a year, concluded
only that in the past the Japanese have not placed great emphasis on original research, not that they cannot do it.[14] Reischauer
has observed that

> A myth has grown up that, unlike other people, the
> Japanese are mere mimics, incapable of invention ...
> In actuality, their isolation has probably forced them
> to invent a greater part of their culture and develop a
> new distinctive set of characteristics [more] than ...
> any comparable people in the world. What distin-
> guishes them is ... skill at learning and adapting.[7]

However, when the Japanese do innovate, they can be frus-
trated. Consider the case of the scientist Junichi Nishizawa. His
first professional achievement, a theory of optical communication, was fiercely opposed by Japanese academicians as an idle
dream. After developing the pin photodiode (1953) and the semiconductor laser (1957), he continued to be mistreated by the academic and industrial communities: "though it was academically
interesting, its practicality was questionable."[15] Even after obtaining a critical patent on a method of utilizing glass fibers for
optical communications in the mid-1960s, Nishizawa continued
to be ignored.[16] But Nishizawa found acceptance in the United
States. He said

> In Japan, scientists labeled a creative idea "impossi-
> ble" right from the start if there was even the slight-
> est technical problem involved. The reason why the
> Americans and English excel is that they relish the
> challenge of tackling difficult projects.

In 1984, thirty years after Nishizawa applied for his first patent, the Japanese academic community paid the price for their arrogance. Nishizawa took his static inductance thyristor, a breakthrough in energy conservation, to the United States rather than suffer the contempt of his colleagues once more. He said

> by bringing this ... thyristor to the United States, I hope to shake up Japanese industry in the good sense of the word. It is a pity that Japanese companies pay more attention to test results from the United States, but cannot use their own judgement about commercializing products. The United States is full of rational, pioneer spirits ready to jump on new products and put them to good use. The difference in environments is what makes or breaks creative research.[15]

Nishizawa is not the only example of a successful, but unappreciated, Japanese scientist. Leo Esaki, one of only three Japanese Nobel Laureates in physics, won his prize in 1973 for semiconductor work while he was employed at Sony. But he accepted a position with IBM because "if you are innovative, you like to challenge the unknown. Unfortunately, the gray nature of Japanese society makes that very difficult."[17]

Nevertheless, despite a reluctance to innovate at the basic research level, the Japanese system is viewed as more successful than that of the West. By way of contrast, Western scientists prefer to follow independent lines of inquiry and expand their own ideas, a process that certainly facilitates creative and innovative thought. In fact, it has contributed to the great Western advantages and strengths in research. But, it has encouraged people to reject the ideas of others. As a result, the "not invented here" (NIH) syndrome prevails in the West. It is common in the West, for example, to have technical groups competing with each other on a particular problem. Each group favors a particular approach and is critical of the other group. Such parochialism even extends into the job opportunities that are offered to graduates. Studying under the wrong professor can seriously retard one's chances for employment in organizations that reject that professor's ideas.

This problem is also manifested in technology transfer where the NIH syndrome impedes the flow of ideas. The impact is particularly felt between research and engineering. It is uncommon in the West for engineering groups to embrace the ideas of alien research groups. Good ideas, therefore, are, at best, deprecated or, at worst, ignored.

Europe, with its nationalistic tendencies, suffers the most from the NIH syndrome. The technology transfer lag there is greater than in the United States. Indeed, as discussed later, even graduate degrees from non-European countries such as the United States are rejected by some European countries on the grounds these degrees fail to meet local standards.

For several reasons the Japanese excel in technology transfer. First, they are skilled at recognizing good ideas (a trait they have demonstrated repeatedly since the Dutch and Portuguese first landed on Japan), and they are prepared to embrace them and incorporate them into their products. Second, the homogeneous nature of Japanese society makes it easy for them to support engineering and manufacturing standards. These standards, in turn, make it possible for the country's projects to share results. Third, the lack of competition in the early developmental phases, supported by the government, works well within that environment and renders it possible to divide up the research problem in the precompetitive phase, eliminating overlap and extracting the most efficiency from the Japanese research community. Fourth, the Japanese reward system is more rational than its Western counterparts. It may consist of power, prestige, and wealth. But, in general, it is not considered in good taste to try to accumulate more than one of those. Greed is frowned upon, for example. This is in contrast to a Western system that drives many individuals to accumulate all three. Thus, Japanese professors bask in their prestige with less need to accumulate wealth, and Japanese corporate officers work for the good of the corporation with no thought of selling stock in the company.

Countries that don't make large research investments, such as Japan, can compete more effectively by getting involved in manufacturing with more cost-effective production techniques. If their labor costs are also low and if there are no international barriers such as tariffs, they can compete favorably with the United States. But advantages change. Japan could lose its advantage to the vigorous Third World or developing nations around the Pacific Rim. Only a few years ago, no one would have expected the Japanese to be buying Korean automobiles or seeking off-shore manufacturing sites in the West. The forces of competition grind inexorably for Japan as well as for America.

Characteristics of Japanese Technology

Because the Japanese are very active in the international sector, they have a great need for automated language translation. Oriental alphabets in general, however, constitute a difficult automation problem for all the Asian countries. The Kanji alphabet (borrowed from the Chinese) has been used to express the Japanese language in written form. With 50,000 characters, it places a difficult technical burden on their computer input, output, and communications activities. This language problem has provided the Japanese with a great incentive to conduct research in related topics such as telecommunications, machine translation and specialized terminals in order to exploit valuable market opportunities in Asia. They have released innovative products in these areas. As a result, in the mid-1980s the United States dropped to second place in the worldwide telecommunications market, with a share of 14 percent following Japan's share of 21 percent.[18]

Nevertheless, if Japan is to prevail in the Technology War, it must find a way to compete with the United States in the strategic issue of software. Japan's historical emphasis on commercial applications must be tempered by an increase in original software development. But, in software, the Japanese suffer a lack of personnel to conduct research across a broad front. Software research is an expensive and demanding field. In the United

States, over thirty years of experience has resulted in the development of management skills, the organization of many teams that satisfy the minimum requirements for size and equipment, a research infrastructure, and so forth. Japan lacks this experience in both research and software.

Research into computer architecture is a strong area in the United States. Nevertheless, the Japanese implementations of marketable IBM-compatible systems are an improvement over the original American designs. So far, this has been a winning strategy for the Japanese, who have found it relatively easy to maintain compatibility with their principal competitor, IBM, in the commercially important data processing markets. Nevertheless, this is becoming increasingly difficult for all suppliers of low-cost compatible equipment as the large American suppliers have become more adept at exploiting the intellectual property rights laws to protect their designs.

Still, Japanese industry is closing the research gap rapidly with American industry in the advanced development of prototype architectures and in the successive refinements of these architectures. Their emphasis on prototyping is largely universal and in contradistinction to Western attitudes. We found a Japanese professor's comments about the relationship between the use of prototypes and the British research program in Advanced Information Processing especially illuminating. He said that the British program would probably fail because it did not require the development of prototypes as do *all* Japanese research programs. We also believe the development of demonstration prototypes is an essential ingredient of technology development.

These forces involving technology and society play a different role in Europe than they do in Japan.

Europe

The Europeans lost their technological edge in the twentieth century because of war and the natural decay that sets in with the collapse of empires. Another less dramatic reason has been their failure to protect their marketshare by not reinvesting their profits in plant modernization and capacity. They have also followed a relentless path of nationalism, which has segmented their marketplace and which has made it impossible for them to achieve the economies of scale necessary for competitive high-technology manufacturing.

The European countries seem to recognize the strategic implications of the high technology markets and place similar emphasis on research and development as does the United States. But their reaction time is slower, a result of splintering among their economic entities and the demanding requirement of building a European consensus. It is also the result of the NIH syndrome and a more entrenched and conservative research and investment community. For example, in the United States, there is a creative and innovative climate in which information technology has thrived.

E. T. Bell, in his wonderful exposition of the lives of great mathematicians, makes our point about the NIH syndrome by reference to distinguished European scientists

> 'Oh, we never read anything the English mathematicians do.'[19]

Nor, as Bell's quote indicates, are British scientists particularly open minded. For example, Britain suffered a lack of support for research in artificial intelligence (AI) for many years that was to some extent due to a serious attack mounted by a prominent British mathematician on a report of the British Science Research Council. As one source put it

> Demonstrating neither understanding nor sympathy, Sir James [Lighthill] declared the work [the 1973 Science Research Council Report on AI] sadly wanting at best and bordering on charlatanism at worst. In either event, it deserved no further support ... a consequence of the Lighthill report was that artificial intelligence sustained a body blow in Great Britain

> ... the superb robotics program at Edinburgh was
> largely dismantled ... For Lighthill had not taken a
> charitable view of the early robotics research. Since
> robotics is about to play a significant role in Japan's
> soaring productivity advances, Lighthill's report was
> a costly one to a nation whose industrial productivity
> is a grim joke.[20]

Funding was essentially cut off for five years because of this attack; five years of some of the most exciting and productive research in AI.

Unfortunately, the United States is not immune from these types of disruption. Work in the United States in machine translation was interrupted after a prestigious National Academy of Science report, published in 1976, suggested that there was little hope in pursuing this complex problem with brute force, that is, computationally intensive methods.

Cooperation versus Competition

There are also European problems in the EEC and NATO alliances over cooperation, standards, and consensus. For example, high technology presents new avenues and demands for cooperation, such as the desire for common standards in networking. Yet, the entrenched bureaucracy holds these opportunities hostage in an attempt to gain leverage in long-standing disagreements, such as those over EEC agricultural policies. Still, the Europeans chafe at criticism over their ability to cooperate. Indeed, a recent ESPRIT meeting theme was "Competitiveness through Cooperation." EEC policy, delineated at the meeting, was to support only these research proposals for EEC funding that would mobilize all the forces of the community at the continental level. We wonder how, in just five years, EEC is going to manage an integration that has eluded Europe since medieval days. The singular successes of the Airbus and the European Space Agency do not demonstrate the massive infrastructure and momentum necessary to compete with the United States and Japan.

Europe is also paying the price for a strong entrenched bureaucracy and centralized powers in the Postal, Telephone, and Telegraph (PTT) authorities. The PTT authorities have, through their political maneuvering, introduced considerable delay in the deployment of new digital switching and standard networking technologies. Although, France, with its Télématique Programs, and Portugal are two of several European nations that have established goals to revolutionize their digital telephone system this century, it is not clear that the French will deregulate their PTT.[21,22] With conflicting interests, competing suppliers, and the traditional European rivalries, there will not be much progress in Europe toward a unified digital network system. With such a network absent, the Europeans will find it difficult, if not impossible, to build integrated information systems and continentwide programming environments that match the sophistication and capability of the United States and the systems under development in Japan.

Sadly, the nations of Western Europe "are nations of eternal war." There has not been much change since Thomas Jefferson wrote those words.[23] As a consequence, Europe has poor prospects of assembling a working consortium and consensus and is lagging badly in the Technology War. Thus, on the main battle fronts, the major combatants are the United States and Japan.

To summarize, we have introduced the major opponents in the Technology War. The United States has a shaky lead but the Japanese are gaining rapidly. The principal factors in the war are not only technology but include economics, education, sociology, and industrial policy as well. In Part II, these systemic issues are explored.

PART II

THE SHAPE OF THE BATTLE

3

The Battlefield

In this chapter, the sciences and technologies that represent the terrain, those critical areas of conflict, are introduced. The competitors that prevail in the Technology War must understand the aspects of these technologies, from basic research to advanced manufacturing. In the Technology War, as in any other war, the terrain determines which maneuvers are possible.

Information technology is one of the great accomplishments that has evolved out of the field of electronics. Electronics was the key, in World War II, to radar, cryptography, and many of the *Wunderwaffen* of Germany. In fact, some British have referred to parts of World War II as the "Wizard War." There is little doubt, indeed, that if conventional wars continue, they will be fought in the electronic arena, albeit with the electronics embodied in modern computer technology. The Soviets refer to this future struggle as the Radio-Electronic Battle.

In any future war, whether it is conventional or not, computers will clearly serve as the basic building blocks of the new weapons. For example, the war between Great Britain and Argentina in the South Atlantic was fought in the electronics spectrum with these high technology weapons. In one battle, HMS *Sheffield* was sunk by a sophisticated French missile. A weakness in the British electronics systems prevented the missile from being detected because the sensors were jammed by a ship-to-shore communication device. Notwithstanding this vulnerability, the British ultimately exploited their technological advantage in the use of modern electronics surveillance technology. They were able to monitor continuously the deployment of the Argentine fleet. Processing this data in real time was a complex task and it required the use of the new information technologies. It was the successful performance of these tasks that led to a victory in the Falklands war.

As the Falklands example demonstrates, the country with the most sophisticated information technology is potentially the strongest military power. Such dependence even extends to nuclear weapons. Information technology and, especially, computers do constitute the major battlefield in the Technology War.

There are several reasons why computers have become so important.They operate at extremely high speeds, have the ability to store and retrieve vast quantities of information, and can make decisions based on the results of previous operations. With just these primitive functions, a generation of programmers has created a body of software that can control a missile, intercept a message, predict the results of an election, or automate a factory.

In the following sections, some of the major areas of information technology are surveyed including hardware and architecture, software, communications, artificial intelligence, and key applications.

Hardware and Architecture

Despite the fact that computers are fast, they are never fast enough. Most computer applications use great quantities of computer time. The benefits that have accrued from faster computers have been offset by a continuing increase in the sophistication of the applications which, in turn, has created an increased demand for computing resources. Thus, increased speed is always desirable.

Speed

Computers do operate at extremely high speeds (about fifty to one hundred million instructions per second, or 50 to 100 MIPS). Specialists use the term MIPS to rate computers just as they use horsepower to rate automobile engines. However, machine speeds are limited by the time it takes their components, such as adders or switches, to communicate and this, in turn, is limited by the time it takes electrons, or signals, to propagate across a wire. One method for obtaining greater speed, therefore, is to build components as close together as possible.

Hardware speeds are also affected by the organization, or *architecture,* of the computer. Von Neumann machines, named after John von Neumann, a mathematician and pioneer in World War II computer developments, are based on a serial design. Only in the cases of supercomputers and special purpose processors has there been any significant deviation from von Neumann architectures. These unique machines offer significant processing improvements over conventional machines for special purpose problems that can take advantage of parallel computation, that is, executing different instructions simultaneously.

Memory

Memory, the medium that holds both data and instructions, comes in two generic forms: high-speed internal storage, typically electronic, and slow-speed external memory, typically magnetic or electromechanical. Naturally, a program that uses high-speed internal memory will process more quickly than one that uses slow-speed external memory. The catch is cost — high-speed memory is typically fifty times more expensive than slow-speed memory.

Increasingly sophisticated applications have led to greater demands for memory, both internal and external. And, as memories increase in size, it becomes more difficult to manage and access them. Since each memory unit must be capable of being referenced (addressed) uniquely, it has become necessary to use larger and larger addresses. These addresses, which are part of an instruction, have required an increased instruction size and a corresponding increase in machine complexity.*

Microelectronics

In the last 20 years, electronics has made great strides. It has become possible to reduce the size of components dramatically. By making everything smaller, and by fabricating entire systems on tiny chips, the new technology of microelectronics has yielded smaller, faster, cheaper, more reliable, and more efficient electronic devices.

Circuits made of microelectronic switches are used to perform logical functions or to store information, which may be either data or instructions. These components are also used to construct larger units that add, multiply, or perform logical functions.

The first generation of computers used vacuum tube technology for switching and logic. These computers were large, slow, and unreliable. They used large quantities of power and needed large cooling systems. In 1947, the industry was ready for the newly invented transistor — a discrete semiconductor device that offered the same function in switching as a vacuum tube but with much better performance in speed, size, reliability, and power. Transistors did not, however, offer great savings in assembly or connection costs. In general, they simply replaced tubes on a one-for-one basis.

The next step in the evolution of microelectronics was the development of integrated circuits (ICs). This resulted in the packaging of larger and larger numbers of transistors and other circuit elements on a silicon chip and therefore reduced assembly costs. Denser packing led to a great reduction in the interconnection distance between transistors. Speed, power, and reliability were all improved. Assembly costs continued to decline as IC

---

*Instructions and data are typically grouped into *bytes*, a unit of information composed of binary digits (bits).

densities increased still more. Indeed, the cost per transistor has been reduced so consistently that, "according to some estimates, a jelly bean (at a cost of one cent) may buy as many as 1,000 bits of memory by 1990."[24] Larger memories are now in production. In 1986, the major Japanese suppliers (Fujitsu, Hitachi, Toshiba, NEC, and Oki Electric) all began mass production of the 1 million bit dynamic random access memory chip, and American manufacturers are following suit. Indeed, NTT recently announced the prototype development of a 16 megabit memory chip.

An important hardware factor that also affects performance is heat. As electrons move through materials there are several effects. Electrical power is used and heat is generated. This heat is a serious factor in hardware design; it limits the effective speed of the device. The heat can be dissipated through a variety of techniques but this becomes more and more difficult as the components are squeezed closer together to attain ever higher speeds. Thus, design considerations for speed and heat are mutually contradictory. It is necessary to balance both on an engineering tightrope. Microelectronics and sophisticated packaging technology have helped.

This tremendous reduction in size through tight packaging and constantly increasing capability is familiar to most technologists. What may not be well understood, however, is the astonishing increase in cost-effective applications that have been realized. For example, while dramatically reducing memory costs and circuit costs, microelectronics helped to create new products and markets, such as the consumer electronics explosion of the 1970s and 1980s, minicomputers in 1965, and personal computers (PCs) in 1980.

Personal Computers and Workstations

Personal computers are machines that, in 1986, sold for under $2,500. They usually include a processor, some high-speed memory (say 512,000 bytes), some disk unit (typically a floppy disk), a video display device, and a printer. A wide variety of options exists for each of the subunits. For example, a printer might be a simple, inexpensive dot matrix device available for a few hundred dollars, or a complex laser printer capable of printing a newsletter or camera-ready copy for a book for $2,500. Personal computer systems range from primitive eight-bit computers to thirty-two bit machines with graphical output.

A personal workstation is a system like a personal computer except that the processor generally has significant computing power (such as a thirty-two bit machine capable of performing several million instructions per second). A workstation with competitive capabilities might be a machine that can execute four MIPS, has 4 million bytes of memory, has a 19-inch bit-mapped screen (which allows graphics applications), a mouse for managing window selections, and networking capabilities. Workstations such as these are designed to run sophisticated software engineering tools and range in price from under $5,000 to over $100,000.

The workstation segment of the computer field is rapidly growing and the technical line between PCs and workstations keeps shifting as the market grows more sophisticated. At the low end of the market, PCs are like commodities with low profit margins, whereas advanced workstations are still much more expensive and offer greater unit profits. Obviously, the producers will continue to develop the technology as rapidly as possible so as to move to, and stay on, the profitable part of the curve. This means that PCs will be continually upgraded with workstation capabilities, and there will be constant pressure on workstation manufacturers to innovate new technology. It is estimated that workstation sales may reach $4.5 billion by 1990.

Supercomputers

The term "supercomputer" is used to refer to the most powerful machines available for numerical computations. Such computers execute more operations per second than any other computer systems. Their speed is measured in millions of floating point or scientific operations per second, MFLOPs. Supercomputers are used for fundamental strategic problems, such as nuclear weapon design, high energy physics, and the simulation of sophisticated weapons systems of the Strategic Defense Initiative as well as for important civilian problems, such as weather forecasting.

Today's supercomputers have provided a new tool for scientific research and development. Originally, physicists had two different methods of inquiry: theory, to provide understanding, and experimentation, to gather empirical data. Now there is a third methodology, *computational science,* which can be added to the basic scientific methodologies. For example, complex aerodynamic structures can be studied under a variety of conditions and even costly wind tunnel experiments can be replaced by simulation models on supercomputers.

Over the last four decades, advances in the performance of supercomputers of about 6 orders of magnitude have been achieved. This dramatic increase in capability has made it possible to apply supercomputers to a broad spectrum of problems and has served as the foundation for production versions of supercomputers.

Improvements in the performance of supercomputers have been based on both the development of faster components and changes in the structure of the systems that allow more operations to be performed at the same time (increased concurrency). Improvements in the applications software have been far less dramatic. Nevertheless, they have taken the form of more efficient computational procedures that use fewer machine steps to achieve the same results. In addition, more appropriate computational procedures have been developed that make use of the hardware in a more efficient fashion.

The United States and Japan have significant presence in the manufacture of supercomputers. Although in the United States during the last decade there has been a lack of development in system software for supercomputers, this has not impeded their use. The typical user of a supercomputer is a scientist or engineer, who is generally familiar with the FORTRAN scientific programming language, and is more willing to assume the burden of creating both application and systems software than is the traditional electronic data processing user.

During this same period the Japanese have shown some real strength in their systems software for supercomputers. The Japanese strength in supercomputer software is unusual since they tend to be otherwise weak and noncompetitive in software. This strength reflects their overriding interest in engineering since supercomputers are excellent engineering tools. Still, despite Japan's strength in supercomputer software, the United States suppliers have dominated the marketplace, probably because the United States is also the largest user of supercomputers.

In fact, most American supercomputers have been designed by one individual, Seymour Cray, who has tremendous respect and credibility in the American marketplace. As a result, his machines are in great demand. To meet this competition, the Japanese, who have not in the past been viewed as a commercial threat by American suppliers, are now pursuing a rational plan to improve the positioning of their machines. This plan includes exploiting their unique upward compatibility with IBM computers (American supercomputers are not upwardly compatible with IBM machines) and establishing a much more aggressive marketing campaign in North America.

Software

It is the programmable and adaptable nature of machines (via their software) that makes them so powerful. Software, which includes all aspects of software engineering, methodology, economics, and maintenance, is as significant as hardware.

Software is the media by which we communicate with computers. For example, we cannot walk up to a computer and state, in simple natural language, a request to perform even the most elementary task. Computers can understand only instructions in *machine language*. It is through software that we convert our desires, expressed in a language that humans can understand, into the machine language that computers can understand.

Machine language comes in many varieties but is basically represented in some numeric form. Software products are available across a spectrum of complexity (or inversely, simplicity of use) to generate those numeric or alphanumeric expressions from languages that humans understand.

The most elementary level of software is an *assembler,* which converts symbolic expressions (such as ADD A) into machine instructions. A slightly more complex software tool, known as a *compiler,* recognizes a more conventional scientific expression such as A+B. An *operating system* is a program that manages the hardware and software subsystems of a computing system. It schedules jobs, allocates memory, input and output, and other resources, and involves assemblers, compilers, loaders, and so forth. It is these systems programs that have simplified the use of computers and dramatically increased our ability to solve problems.

Applications software (such as accounting systems, word processing systems, and spread sheets) often are developed for specific hardware and a specific operating system. In most cases, the software is restricted to the parent hardware and software and operating systems configuration under which it was developed, because the ultimate machine language will operate only on that configuration. Although there has been some progress in the establishment of standards in programming systems, the industry still suffers a serious disadvantage in that software investments are not readily portable from one machine to another. Competitors that can find a way to port machine language from one hardware and software configuration to another can have a significant cost advantage over their rivals.

Finally, software maintenance costs depend primarily on the quality and reliability of the original software. It is, perhaps, from 10 to 100 times more expensive to repair a program error than to make a smaller up-front investment in improved design and programming and thereby avoid the error. Thus, any competitor that can produce more reliable programs gains a significant competitive advantage.

Computing and Communications

Communications is a natural partner to computing. Indeed, many information processing functions are impossible to distinguish from communications functions. For example, a data processing problem with geographically distributed databases may make heavy use of communications. A credit card used in California may be verified by inserting it in a special reading device that calls a local computer, which, in turn, calls another computer in New York that provides the final verification.

Furthermore, performance, or throughput of the overall system, may be actually more dependent on communications, network architecture, or communication device technology than it is on computer processing. Similarly, the hardware technology and software that is embedded in communication systems, from Department of Defense packet radios to AT&T telephone switches, make heavy use of new computer technology. Because digital technology is also more reliable, more accurate, and cheaper than conventional telephone technology, there is also an ongoing evolution from the older analog-based hardware to digital-based networks. This further facilitates the marriage of computing and communications since they are based on the same hardware processes. In some cases, even the same software may be employed for both processes.

Artificial Intelligence

Researchers into artificial intelligence have forecast exciting new applications of computers that are of enormous economic consequence. These range from medical diagnosis, the ability to function at the expert level of competence, to the use of natural language for use with computers. Thus, this "New Frontier" is attracting the attention of all of the technology warriors.

As the information society evolves, the next source of fundamental change could be the applications of artificial intelligence technology. Once considered too exotic ever to become a reality, let alone a commercial opportunity, AI is now showing some evidence that it may have useful and practical applications. AI may offer commercially viable solutions to current problems that have not been attacked in the past because of the inadequacies of conventional data processing technologies. In defense problems, AI may be especially useful in a myriad of military applications including command and control, intelligence, and the Strategic Defense Initiative. It offers especially interesting opportunities since it suggests autonomous, intelligent, and self-functioning weapons are possible.

Any definition of artificial intelligence is likely to be controversial among those working in the field. A part of the problem of definition lies in the changing nature of AI — what was considered complex AI technology twenty-five years ago is considered routine computing today.

Experts disagree as to whether it is possible to label machines "intelligent." The *Turing Test*, developed by Alan Turing in 1937, stated that a machine would have "machine intelligence" if a human observer could not distinguish the response of the machine from that of a human being. This test has continued to be the standard by which advances in AI technology are measured. As evidence of the increasing sophistication of computers, it might be interesting to observe that at the 1985 Annual Meeting of the Association for Computing Machinery, several human observers were *unable* to decide whether a computer or a person had made particular chess moves.

We are not suggesting that what happened at the meeting implies that machines can think. Computers do not exhibit intelligence except in some very limited and unimpressive ways. The question of whether machines can think may have some philosophical value, but it is in itself unlikely to affect the outcome to the Technology War. Nevertheless, the issue continues to run hot and cold:

> [Artificial] *Intelligence* was a political term defined by whomever was in charge. This accounted for its astonishing elasticity ... the question — Can a machine

think? became once more a nonquestion, a nonissue
of no consequence ... while we weren't looking, the
burning question of a decade or so ago — Can a machine think? — turned from white heat to white ash
... human vanity, not human science was the real
issue.[20]

Still, thinking machines are the Holy Grail of AI research.
Although we cannot expect very useful results from the current
research efforts, it is a fascinating concept. It will undoubtedly
motivate more investment in the research topics that lie at the
foundation of the Technology War, expanding the base and offering interesting topics for speculation in the twenty-first century.

Besides the quest for thinking machines, researchers in AI
have sought better ways to communicate with machines. There
is, for example, considerable interest in systems that use a form
of everyday English to query databases. It is believed that such
natural language systems will make it possible for a much larger
segment of the population to interact with computers. Indeed,
people who may be unable to learn or who are not interested
in learning complex programming languages should constitute a
large market for expanding data processing applications.

One interesting and related application of natural language is
the machine translation of languages. The objective of machine
translation is to replace all or part of the human effort in text
translation by machine processing. Current products range from
on-line terminology data banks (dictionaries), to partial translation systems that are integrated into word processing systems, to
fully automatic systems. The translation market, currently satisfied by human needs, is an attractive target. In fact, in Europe,
university degrees in translation are in great demand. Even partial automation of the process could encourage significant growth
in this market. Skilled translators are hard to find, many translation jobs are tedious, and high costs and delays discourage the
use of existing translation services. With the handicap of a non-
Western alphabet, the Japanese especially feel an attraction to
machine translation. As a major trading nation, they need to

have efficient means of getting information to and from their markets. The Japanese, therefore, have placed great emphasis on the development of automatic machine translation systems and are considered to be the leaders in the field.

Although thinking machines are not on the horizon, some very limited forms of thinking-like processes have been developed. Expert or knowledge-based systems are computer programs that reason, or make inferences, from sets of facts that may be incomplete, inconsistent, or uncertain. These systems try to codify facts that have been extracted from human experts. They are programmed to deal with the problem of inference and to mimic some level of human reasoning. They usually operate on a limited domain of expert knowledge, and their technology is still largely experimental.

Knowledge-based systems represent perhaps the most well known and aggressively pursued area of AI activity. They provide some systematic capabilities for capturing, refining, packaging, and distributing expertise from more than one person and for solving problems that are too complex for large numbers of untrained humans. One of their corollary benefits is that in the development of the knowledge-based system, the process forces people to codify their knowledge and rules. Indeed, it is common in such processes to discover great gaps and differences in what employees and companies think are their common policies and procedures.

In early 1980, there were only a few knowledge-based systems. By the late 1980s, hundreds are under development. Some of the most popular application areas are medical diagnosis and treatment, manufacturing and configuration planning, image understanding, military threat assessment, mineral exploration, and financial services advising. These application areas represent niches that cut across industries and domains. For example, intelligent instruments have applications ranging from consumer to military to medical use and domains ranging from electronics to sensing to hematology.

PROSPECTOR is an experimental expert program that helps

geologists evaluate the mineral potential of an exploration site or region. It consists of a number of models of particular types of ore deposits, coupled with a reasoning mechanism and modules that conduct a dialogue with a user who has information about a particular exploration. Each model is obtained from an exploration geologist who is a recognized authority on the type of deposit being modeled. PROSPECTOR currently contains models for several classes of copper, molybdenum, uranium, and other deposits. It is given credit for locating a molybdenum deposit worth $100 million. Nevertheless, the deposits identified by PROSPECTOR were located in a field in which molybdenum had already been found and the ore was actually deemed too deep to be worth recovering on an economical basis.[25]

Despite the hype, there are still few successful knowledge-based systems in use today. Most real applications demand more capability than is available in a cost-effective manner and the existing expert systems are quite limited in their range of applications. For example, Professor John McCarthy reported at the 1985 United States–Japan Seminar on Knowledge Systems that a famous medical diagnosis expert system developed at Stanford University, actually "understood" less biology than a high school student. If certain recommended treatments prescribed by the system were followed, he said, the patient would probably die.

Despite appearances, knowledge-based systems do not behave like human experts. They do not learn from experience nor do they understand the problems. They are unable to generate expertise, and their knowledge must be given to them. For this latter reason it is impossible to create a local expert system without a human expert, a fact frequently overlooked by zealous developers anxious to develop knowledge-based systems to replace missing human experts.

Applications

Along with the conventional data processing applications, such as general ledger, accounting systems, orbital mechanics, and inventory control, there is a series of high technology applications especially related to the engineering and manufacturing technologies. These applications are at the forward edge of battle in the Technology War because of the role they play in the competitiveness and productivity of the combatants. They include computer-aided design (CAD), computer-aided engineering, computer-aided manufacturing (CAM), computer-integrated manufacturing (CIM), robotics, flexible manufacturing systems (FMS) and mechatronics — a phrase coined by the Japanese to describe the union of mechanical and electronic engineering. Mechatronics includes such diverse topics as flexible manufacturing systems, vision systems, assembly and inspection systems, intelligent mechanisms, software, manipulators, and precision mechanisms. Both CIM and mechatronics are needed to produce the next generation of machines, robots, and smart mechanisms for applications such as manufacturing, large-scale construction, and work in hazardous environments.[26]

CAD is an interactive process whereby an engineer works at a video display to capture and interact with a geometrical database and other engineering tools, such as simulation packages and sophisticated graphics. The objective is to provide a design facility that allows the engineer to experiment with a multitude of designs and to study the performance characteristics of the designs. One example is designing air foils for wings for commercial aircraft. With today's systems, engineers can enter wing data, create graphic illustrations with three-dimensional views, simulate airflow, lift and drag, change parameters, and study the results in real time. In electronics design, engineers can use the same type of facility to study different circuit layouts and routings.

Computer-aided engineering is the use of analytical software, much like simulation in the above example, to analyze the design characteristics of a machine part or device. It includes such tools as continuous dynamic simulation, finite element modeling, thermodynamic stress models, deformation, and many other kinds of engineering analysis.

Computer-aided process planning is the use of computers

to help plan the operations that each machine part would go through on the factory floor, that is, selecting machines, tools, and sequences. It is especially useful in material planning, distribution and inventory problems.

CAM is a widely used term with various meanings. It is most commonly defined as the application of computer equipment in converting design information into an actual product. This includes the generation of control information and instructions to run numerically controlled machine tools, robots, FMS, automated materials handling systems, process controllers, programmable controllers, computer-aided inspection and testing machinery, and automatic assembly systems.

Numerical control (NC) technology was the first instance of CAM. In NC, a part programmer, usually a person with a machinist's background, "programs" a part design in a geometric language that is then used by a computer to drive the machine tool through the sequences necessary to machine the part. Thus, an aircraft manufacturer need not store the part in inventory; a Boeing 737 wing, for example, need only be stored in the form of a paper tape for a 100-foot skin mill at the Wichita plant of Boeing Aircraft Corporation. NC is truly the first step in a CAM process and has proven cost effective even in instances of production runs of only *one* part.

The Japanese have certainly recognized the productivity improvements that come from NC and the market demand for it. In just seven years, they captured 50 percent of the NC machine tool marketplace in the United States. In fact, imported machine tools now constitute 50 percent of the *entire* machine tool market in the United States.

Beyond NC, the ultimate in manufacturing automation is represented by FMS. It is a combination of several machine tools, a material-handling system, and a central computer controller to direct work and machine use.

Computer-aided inspection (CAI) is yet another use of computerized engineering design data in manufacturing, this time in the quality control area. In CAI, coordinate measuring machines are controlled by computers that rely on data from a company's engineering design database, to measure parts automatically and ensure that they have been manufactured to the specified design tolerances.

One segment of CIM is industrial robots. These are computer-controlled devices that automatically perform a programmed sequence of operations. The Robot Industries of America defines a robot as a "reprogrammable multifunction manipulator designed to move parts, materials, tools, or specialized devices through variable programmed motions, for the performance of a variety of tasks." Western Europe has a similar robot definition, but the Japanese robot definition is broader, including manual manipulators and fixed sequence machines.

Some of the benefits of using robots include lower manufacturing cost, improved product quality, and improved working environment. Lower manufacturing cost occurs as a result of reduction in labor costs, higher output from the same floor space, lower materials costs due to better material usage and less scrap, less working capital in inventory, increased control over critical aspects of the manufacturing process, and simplified management of the manufacturing process.

Japan, of course, is renowned for its use of robots and automated factories. Indeed, Japan's robot population is numbered in the hundreds of thousands in contrast to the United States and Europe where they are numbered in the tens of thousands. Nevertheless, many of the Japanese robots are simple devices using only a few degrees of freedom and are suitable for only one task (or one set of tasks). In the West, on the other hand, it is more common to use sophisticated and programmable robots. As a result, the United States tends to concentrate on the development of new robot technology, while the Japanese continue to place their emphasis on today's manufacturing problems with simple, cheap robots.

The sciences and technologies we have discussed, from their origin in basic research to their exploitation in advanced manufacturing technologies, set the stage for the Technology War.

In the Technology War, just as in any other war, this terrain is critical. What makes this terrain especially interesting is that in maneuvering across it, we involve the transfer of technology from one stage (and owner) to another stage (and another owner). This process of technology transfer is just as important as the development of the technology.

4

The Flow of Technology

Technology transfer between two parties is most effective when there is close trust and cooperation. That necessitates a free and open exchange of information. When a Technology War occurs, however, the same things happen to technology transfer that happen to free speech during an ordinary war. One needs to communicate intelligence information (proprietary data) over a secure channel within the army (or company), but barriers must be erected to keep the enemy (or the competition) ignorant of strategies, intentions, resources, the deployment of forces, and so forth. Technology transfer, a terribly complex process, becomes an awesome management challenge.

Technology Transfer

Improving the flow of technology from research into products is essential to winning the Technology War. Accomplishing that means dealing with a fundamental management problem, the complex task of finding a method for shepherding an idea from discovery to application.[6] Managing the transfer of technology poses difficult problems within a single laboratory site, a company, or a nation. The complexities increase in unlimited ways as the technology crosses international boundaries: the relevant factors include education, sociology, economics, government policies, law, and national security.

Science versus Technology

To grasp the transfer mechanisms of technology, we need to begin with an understanding of the differences between science and technology. *Science* deals with the understanding of both the nature and properties of the universe and the laws or theories that govern or explain physical phenomena. *Technology* is applied science, although it is sometimes difficult to distinguish it from pure science. At the extreme, however, it is simple to separate very theoretical science from very applied work. This point can be illustrated using nuclear physics and the atomic bomb. For example, it is easy enough to accept as pure science the theoretical work of Albert Einstein, illustrated by his famous equation,

$$E = mc^2$$

which relates energy E to mass m in terms of the velocity of light c. On the other hand, the designers who created a fissionable weapon based on this science were clearly doing technological work. But these classifications break down as we examine the relationship in greater detail.

The Einstein equation suggests that matter can be destroyed only with the release of large quantities of energy. By 1939, it was understood how the uranium atom could be split into the atoms of different elements with a loss of mass. This, according to Einstein, required a corresponding release of energy, which suggested uranium as a suitable material for a fission bomb. All laboratory experiments at that time worked with U^{235}, an isotope of uranium that comprises about 1 percent of the natural ore, the one found most commonly in nature. U^{235} fissions easily, whereas, the prevalent isotope, U^{238}, does not. For the process to work, kilograms of U^{235} were needed, but only tiny amounts had ever been available in laboratories. Searching for other materials for a bomb, it was discovered we could transmute (that is, convert)

U^{238} into plutonium, an element that does not appear in nature but that is highly fissionable. The trick, then, was to separate the plutonium from the uranium by a chemical or physical process. In early 1941, the plutonium isotope, Pu^{239}, was produced at Berkeley and was fissionable as predicted. But how was one to separate the desired material from the U^{238}? Weapons could be made with either fissionable material, but both substances were in short supply.

There were two processes invented for separation of U^{235} and plutonium from U^{238}. One was an electromagnetic and the other a gaseous diffusion process. The diffusion process, demonstrated early in World War II, pumped uranium hexafluoride gas, an extremely corrosive substance, against a porous barrier. Lighter molecules, which contained U^{235}, passed through the barrier more rapidly than the U^{238} molecules. The gas cycled through a large series of stages and became progressively richer in U^{235}. Unfortunately, the yield, that is the increase in the density of U^{235} at each stage, was so low that thousands of stages were found to be necessary. Since each stage comprised highly complex chemical engineering processes and needed facilities to deal with the corrosive nature of the gas, thousands of man hours of engineering were needed to realize deliverable weapons. After four years of hard technology transfer work, the gaseous diffusion and electromagnetic techniques were combined to produce the quantities of fissionable material that were needed for the early bombs.

It is interesting that up until 1942, the Germans had kept pace in basic nuclear research with their colleagues in Europe and the United States. Thereafter, because of a breakdown in their experimental innovation, they fell behind at the rate of one year for every month.[27] In a major oversight, they failed to discover that carbon could serve as a moderator (controller) for a nuclear pile and they relied exclusively on heavy water available, at that time, only in Norway — a simple target for Allied bombers and special forces. Fortunately for the Allies, no one *transferred* the fact that carbon was a useful material to Germany.

In our example, we started from theoretical insights expressed in a simple and elegant form. But, at each technology phase, increasingly complex descriptions were required. One aspect of the technology transfer process is clear: Pure science is easy to transfer, and applied technology is hard to transfer.

Managing Technology Transfer

Managing technology transfer is made even more complicated if the government is involved in the process. When a company has had a major breakthrough and needs to transfer the new technology to manufacturing, the process seems to take forever. In pharmaceuticals, for example, the Food and Drug Administration testing process routinely takes up to ten years. Further, technology transfer can be accidentally or deliberately impeded by the government when it establishes rules, guidelines, and definitions regarding a new technology. For example, the U.S. government recently promulgated a series of rules that determine how five U.S. federal agencies will deal with biotechnology and genetic engineering. These rules presumably establish a system to protect the public, but they are certain to slow the spread of the technology. These rules have resulted in a biotechnology battle in the United States, where some people demand faster development of biotechnology whereas others deplore the current pace of innovation.

The cellular telephone is another example in which government regulation and procedure delayed the flow of technology to the marketplace. In this case, it took the FCC seven years to issue operating licenses to cellular telephone operators. During this same period, the Japanese deployed and field tested their systems in Japan. The result was superbly engineered Japanese cellular telephones that captured the American marketplace when it was finally opened for service.

On another front, the 1965 Brooks bill and its successor, the 1982 Paperwork Reduction Act, introduced an administrative swamp that delayed the government itself from using state of the art technology and imposed on each federal agency the requirement to institute complex bureaucratic systems to monitor and manage their data processing systems. Other government acts of some consequence came in 1968, with the United States versus IBM antitrust suit and also the consent decree imposed on AT&T in their case with the government. Both of these cases throttled the pace of innovation; for example, IBM's generosity toward universities during the 1970s was severely curtailed and AT&T was compelled to stay out of the computer business for some time.

The Privacy Act, Presidential Directives, and other bills, from 1974 to date, have been useful in protecting citizens' rights and imposing more secure communications and computing systems on federal users. But, along with the consequences of the Brooks bill, the cost in reduced technology transfer has been enormous.

In the commercial sector, when a company succeeds in making a breakthrough, technology transfer to its competitors seems unusually quick. There are numerous tales about products based on ideas described in conference papers, entering the marketplace before the idea is disseminated to branches of the same company that presented the paper.

Obviously, technology transfer is undesirable when research results from a company's laboratory are exploited in its competitor's products. A good example occurred with products that compete with those of the Xerox Corporation. During the 1970s, the Xerox Palo Alto Research Center (PARC) was a leading software research center. Important innovations in programming languages, networking, laser printers, and the development of workstations were accomplished there. But the product ideas developed in PARC were not so attractive to upper management. When Xerox applied the same financial guidelines used for copier technology to computing products, they could not pass the internal Xerox profitability standards. Thus, when the Xerox computers, such as the STAR, were available, they were not particularly cost-effective, were not marketed aggressively, and, consequently were not successful commercially. This demoralized the staff at PARC, and some left for other companies, most notably Apple, where Xerox innovations were exploited much more successfully in products such as the Macintosh. In this example, Xerox not only failed to exploit its own technology, it provided the training ground for more entrepreneurial inventors elsewhere.

IBM suffers from similar problems in technology transfer. For example, it has been said of IBM

> ninety percent of those bright ideas somehow just lay around without ever being developed. IBM is a corporation with its own technology transfer problems from research to development.[20]

Moving technology into production is neither easy nor automatic, as our Xerox example showed. There is no simple check list. There are many obstacles; technology transfer is not a natural process. Often, researchers who invent new ideas will not take the time to work out details at an applied level. People with good communication skills are needed to serve as a bridge with the researchers and production engineers. Spencer, who experienced these agonies as the Xerox PARC Center Manager, has stated

> Constant efforts [for technology transfer] on the part of managers, researchers, manufacturing and marketing organizations are required. The establishment of good communications and trust are essential.[28]

Geographical dispersion doesn't help. In most major companies, it is almost impossible to have all the people involved in a complex product located at the same facility; in multinational companies it is impossible. Many companies perform research in one location, whereas their production people might be at an overseas plant to exploit lower labor costs. In addition, for high technology products, marketing people must be in the country of sale. There is no other way that products and their critical sales materials can be tailored to the unique needs of the target markets. Transferring the technology to these people is an awkward and tedious process.

Encouraging Technology Transfer

Technology transfer can sometimes be induced under the right circumstances. The Israelis, for example, were literally pushed into the development of fighter technology. Political events that threatened the state's existence persuaded Israel that its suppliers of planes and engines were not dependable. To develop an independent capability, technology was licensed from the United States, acquired by other means from France, and developed domestically to meet the need. Today, the Israeli air force is highly respected for both its equipment and tactics. And, because Israel can point toward technology honed in actual combat, its aircraft now are in great demand. Israel today exports the cost-effective Kfir and other medium performance fighters that they were forced to create. Indeed, Israel is now one of the world's ten largest exporters of arms.

In another approach to this problem of technology transfer, the West German government is proposing the formation of an International Computer Science Institute to be located adjacent to the campus of the University of California at Berkeley. In return for five years of support, the German government plans to send a specified number of scientists and engineers to participate in the Institute for several years. The researchers would then return to Germany taking new American-based technology with them. This rotation of researchers is a relatively new phenomenon, having been instituted by the Japanese in their fifth generation research programs. Studying overseas, however, is an ancient method for learning from your adversaries. This plan to study at another country's university is the reverse of a situation that existed in Germany. Between the world wars, for example, Germany led the world in "modern physics," and a postdoctoral fellowship at a German university was a sine qua non for a career in physics.

Hiring foreign consultants is another way to transfer certain kinds of vital information. In general, consultants offer an efficient and inexpensive method for transferring up-to-date information. The Japanese excel at extracting information from foreign consultants because they are well motivated and prepare assiduously for each session. This motivation is not a twentieth century phenomenon, the Japanese have always been quick to recognize the value of foreign ideas. After the Meiji Restoration, for example, they sought to acquire the military and economic technology of the West. At that time, the Dutch were the principal source of such information, and Western skills came to be known as Dutch Learning. In order to encourage the development of those skills, the Japanese rewarded persons knowledgeable in these "barbarian" skills.[7] Contrast this with the attitude in Britain where engineers are demeaned.

A good deal of information about high technology can also be acquired by reading the open literature. In fact, the Press has always been a notorious source for technology intelligence, if not technology transfer. It is well known that

> the cheapest and fullest of all sources of information has always been the Press. So before the war [World War II] everybody used it ... reading the general and technical press of their target countries.[27]

The absence of material in the Press can also trigger technology transfer. For example, in 1940 the sudden lack of papers on nuclear physics in the United States and United Kingdom alerted the Germans and Soviets that work of some consequence was underway in the West. Illuminating, perhaps, is Reischauer's observation that the Japanese "probably receive more world news than any other people."[7]

Intellectual Property Laws

As part of their national policy, governments enact protection laws so that inventors and innovators are encouraged to share their ideas in order to improve the flow of technology and still enjoy some economic reward for the use of their ideas. In the

United States, the enabling authority is the U.S. Constitution (Article I, Section 8):

> To promote the progress of science and useful arts, by securing for limited times to authors and inventors the exclusive right to their respective writings and discoveries.

From this section of the Constitution has sprung the trademark, trade secret, patent, and copyright laws that are intended, in one form or another, to protect inventors and motivate them to share their ideas while simultaneously rewarding them with economic gain.

These protection laws, known as *intellectual property laws,* protect different aspects or uses of ideas or inventions. They are intended to spur the transfer of technology and especially its use; however, the laws can, and do, restrain or restrict the flow of technology. Because the intellectual property laws vary from state to state in the United States and, more extremely, from country to country, it is difficult to anticipate in any given case whether they will accelerate or impede the transfer of technology. Copyrights, which seem to be the most applicable to computer software of all the intellectual property laws, are typically long lived, protect the expressions of ideas (but not the ideas themselves), and are applied uniformly across the United States and, to some extent, internationally.

There is some confusion as to how copyright laws affect both programs written in high level languages and their corresponding machine language versions. Nevertheless, under the 1980 amendments to the 1976 Copyright Act, protection of machine language from copying and duplication is stabilizing. Still, there continue to be vast areas of unresolved law or inconsistent case law concerning software written in high-level programming language and its subsequent intermediate steps to machine language, the use of new media for storing software, the impact of standards, and the use of software libraries. The lack of commonly accepted legal principles, skilled judges, and lawyers will undoubtedly lead to years of thrashing in the courts. Most of it will focus on software and its representation of ideas.

It is interesting that Apple's LISA and MAC products were inspired by ideas developed at Xerox PARC. However, Apple was more sensitive to the value of these ideas, such as the user interface, and broke new legal ground with its actions against others who it claimed were infringing on *their* software environment.

Because software is an increasingly valuable asset, invoking such protection is a growing practice on the part of many software suppliers in the United States. It is expensive and does require the manipulation of complex law. It is no simple trick to separate an idea from its expression in a computer program; after all, we must have some language in which to express an idea. It is even more difficult to distinguish between an algorithm or process and its embodiment in a computer system.

With all of this activity, attorneys and the courts will face an expanding sea of uncertainty about the scope of copyright and trade secret protection, sadly at the expense of the free interchange of ideas and free competition. Case law will continue to be developed on an ad hoc basis, probably without regard to technology, as has largely been true in the past. This will have an unclear impact on the transfer of technology embodied in algorithms.

Patents, which are used to protect hardware, algorithms, and processes, also are coming into wider use for the protection of software. Patent protection suffers, though, by exposing the ideas that trade secrets protect. The inventions are also subject to the Patent Secrecy act, which allows a government agency to instruct the Patent Office to place a confidentiality order on an invention and proscribe its use (in these cases the inventor is entitled to "adequate" compensation). These secrecy orders are not used frequently (the Public Cryptography Study Group reported only seven applied to cryptology in 1981) and are applied usually to patent applications that have implications for weapon systems or cryptography.[29] With the increasing military dependence on computing, there may be more secrecy orders.

The Problem of Piracy

Protection mechanisms affect the lifeblood of businesses that depend on technology transfer. Many a new high technology company has managed to obtain its venture capital financing based on patents, copyrights, or trade secrets — or the absence of them in the marketplace. Many joint ventures or limited research partnerships have been founded to develop or exploit these intellectual property rights.

The protection mechanisms are intended to reward the clever innovator through financial gain. They offer the inventor a competitive edge in his market for some period of time. To overcome that edge, many forms of reverse engineering (to extract the ideas, concepts, or inventions) are used. Of course, determined competitors will also seek information through the more traditional forms of industrial espionage (such as stealing the source code or documentation). Most developers of software are relatively casual about their innovations and few protect them. It is relatively simple to gain access to the source code.

From the perspective of U.S. suppliers, there are countries whose compliance with the Universal Copyright Convention (or other covenants such as the Berne Convention) is inadequate, and the costs to the United States for such noncompliance can be high.[30] The Department of Commerce estimated that at least 750,000 U.S. jobs may have been lost because of counterfeit high technology products. The People's Republic of China, Hong Kong, Taiwan, Singapore and Indonesia are countries where, in the past, piracy has been considered especially rampant. Even though such practices are typically directed at conventional publishing products, and off-the-shelf products, rather than leading-edge technology, they offer the pirates opportunities to reduce lead times drastically, to exploit their cheap local labor, and to evade the tremendous software investments of the original developer. These practices clearly interfere with the United States' ability to remain competitive in the international and domestic marketplace through innovation.

There are, however, some promising signs. In Singapore, a state noted for its ability to innovate, there have been strong government recommendations to establish better means for the protection of software. Under the leadership of the National Computer Board, a committee on the Legal Protection of Computer Software was established with representatives from the Attorney General's office, Department of Education, and so forth. Among their major recommendations were proposals to amend the copyright laws in Singapore to protect software and a strong recommendation to take a public stand against software piracy.[31]

The intellectual property laws also may be used occasionally to protect comparative advantages and local industries. Japan, for example, has entertained legislation that would have allowed MITI to enforce compulsory registration and direct the granting of one firm's software licenses to other competitive suppliers when MITI felt that such action was in the national interest. There were some particularly onerous and complex aspects of this legislation with respect to what are known as derivative works (altered from the original). The Japanese characterized compulsory registration as follows:

> in order to work toward more efficient usage of programs, use licensing of existing programs ... in the same manner ... with respect to rights concerning programs as well ... in cases where it is necessary in the public interest, or where there is non-working of a program, it is necessary to establish a licensing system for use and reproduction, and so forth, by means of arbitral [compulsory] procedures.[32]

Such a mechanism would, if mandated by law, effectively place the authority to make and distribute copies of software in the hands of a governmental Program Review Committee.[32] The government's proposed authority to issue copies of valuable software to Japanese suppliers was quite objectionable.

The Japanese recognize that the copyright of software is one of the effective protective mechanisms available to them.[33] Other legal means are less effective in Japan, because the law is limited by "the fact that the Japanese people have a primitive or immature conscientiousness of contract."[33] It is common in Japan to commence negotiations *after* an agreement is reached. Therefore, the Japanese often question the need for written agreements because they feel that their loyalty or trust is in question. Even when the contract is placed in written form, the Japanese simply consider it a starting point for future discussions. They do not consider it particularly binding on them, and they may not even read the contract. This is the result of the application of Chinese traditions of reliance on moral rather than legal principles. Historically, most Asian relationships have been between a subordinate and his leader and they implied unlimited loyalty, not a legal relationship. Law is simply alien to the Japanese.[7] Their consciousness is being revised, however, as they increase their interactions with the West.

With a worldwide legal and political system mired in nineteenth century traditions and with twenty-first century technology on the store shelves, we must be cautious: The ideas may transfer much more quickly than expected or the laws or policies may interfere with the commercialization of the ideas. Alternatively, there may be unexpected breakthroughs in technology or in the time expected to develop the technology. The nimble warrior will find the best way to exploit these systemic advantages and disadvantages.

Export Controls

In the United States, the Department of Commerce manages the flow of technology out of the country by regulating the export of all computer-related technology. This includes software and, most notably, the leading technologies of the fifth generation.[30,34] Actually, nothing manufactured in the United States, with the exception of foodstuffs, can be exported without government authorization. The authority for this control derives from the Export Administration Act. The amount controlled is awesome. A recent National Academy of Science study suggested that over 40 percent of American manufacturing exports in 1985 (about $65 billion) were subject to these or other more stringent reviews.[35]

Controlled items are documented in the Commodity Control List, which designates whether an item may be exported under a *general license* (essentially no restriction) or a *validated license* (differing restrictions depending on the item and its destination).

Computers and associated peripherals, control software, and their related technical data, such as documentation, specifications, and manufacturing data, are all controlled. General purpose and applications software is frequently considered technical data under the Export Administration Act and it requires licenses when shipped independent of the hardware. Clearly, the government can and does encourage (or discourage) the flow of technology by the manner in which it maintains the Commodity Control List or, indeed, the speed with which it processes (or does not process) applications for licenses.

Governments can use these techniques to control the flow of technology to allies as well as adversaries. In the United States, there is a preoccupation with national security that extends to the flow of information to allies because of concerns that the technology will be reexported to adversary nations. This preoccupation with the specter of Soviet intelligence activities, frequently interferes with the government interest in encouraging overseas computer sales to improve the balance of payments.[36] The result is a schizophrenic approach to export control, compounded by bureaucratic insensitivity, which infuriates scientists, vendors and users. Such inconsistent behavior has been demonstrated clearly in the Iran–Contra arms scandals. Curiously, it even extends to the level of IBM-compatible personal computers in which there is considerable doubt as to whether the machines should be restricted at all.[37]

Technology for military or nuclear facilities use is strictly controlled by Department of State and Defense agencies. Such technology, including software, may fall on the U.S. Munitions List or on the Military Critical Technologies List. Since portions of the

Military Critical Technologies List may be classified secret, some vendors can violate the law without the benefit of any advance public notice. Some importers suspiciously view the classification of the list as a U.S. government scheme to circumvent the trade treaties and restrict the flow of technology for competitive commercial reasons; that is, just as a U.S. maneuver in the Technology War.

Under the Arms Export Control Act, the President is authorized to issue the International Traffic in Arms Regulation that identifies specific types of articles that, for export, require a license granted by the secretary of state. We frequently find these controls applied to cryptographic devices.[29] With an increasing use of encryption, electronic funds transfer, and protected commercial telecommunications however, many devices that otherwise might not be considered cryptographic devices, such as banking terminals, can fall under these controls.

Furthermore, the International Traffic in Arms Regulation has been used by the government to slam the door in the face of foreign scientists desiring to attend conferences in the United States. Under the International Traffic in Arms Regulation U.S. scientists cannot publish certain technical results abroad, or even disclose technical data at a meeting in the United States with foreign scientists present before the material has been published domestically. This means that export licenses may be necessary for foreign employees working in U.S. laboratories.

Supplementing the International Traffic in Arms Regulation is the Commodity Control List. Items on this list may not be exported to communist countries. The management of this list is administered by the Coordinating Committee, which is composed of most of the nations in the Western alliance and Japan and prepares lists of items that would be deemed dangerous to the alliance if the items were exported to Soviet bloc countries.[34] The Coordinating Committee's main objective is slowing the flow of technology to adversary nations. Virtually all the interested federal agencies — state, commerce, defense, CIA, energy, and NASA — get involved in selecting Commodity Control List items.

The rules of the Coordinating Committee and the other regulations are also used to control the reexport of U.S. technology. Since the export control laws have extraterritorial force, U.S. companies are expected to enforce these regulations even among their subsidiaries that may be operating under different foreign laws. Enforcement can get quite complicated among multinational companies. Still, such delays introduced into the transfer of technology by these processes are short lived.

Sometimes export regulations are contrary to the economic or defense interests of a country. For example, current U.S. Defense Department policy restricting the export of dual-use technologies may encourage European firms to enter alliances with Japan or other producers who they would otherwise prefer to avoid. See, for example, our discussion about the British Alvey Program in Chapter 10. Also, because of U.S. defense restrictions, some European firms cannot rely, with reasonable assurance, upon the fact that the technologies that are needed for their final systems' products will be supplied by the United States. As European nations turn to other sources, American firms lose sales opportunities. President Reagan's Commission on Industrial Competitiveness suggested that the annual loss in sales to American firms, as a result of export control problems such as these, could be as high as $12 billion. This explains the U.S. Senate's proposal that would deny American markets to any countries that might take advantage of U.S. government sanctions. Such sanctions would deny American companies access to their existing markets, such as South Africa and Libya. The Senate is loathe to see foreign competitors exploiting these American restrictions. Short-term defense and economic policy, thus, can work against the very industrial sectors that are fundamental to the long term defense and economic interests of the country.

Unfortunately, in restricting the flow of technology, science suffers as do some constitutionally guaranteed liberties. Whether the restriction is for national security reasons, for competitive advantage such as protecting the proprietary rights of an invention, or protecting a comparative advantage, such as American expertise in software technology, the free interchange of information is impeded. The management of technology transfer for competitive or strategic advantage is thus a double-edged sword. In managing it, we need to use all the weapons at our disposal: good communications between scientists and engineers, enlightened and informed domestic government policies and regulations, a rational judicial interpretation of the intellectual property rights laws, and a foreign trade policy that optimizes the national strategic interests of the United States.

Technology Transfer Goes to War

In the present Technology War, it is not just the United States against Japan or other international foes, but companies like DEC versus Hewlett-Packard and IBM versus AT&T, and the city of Austin, Texas, versus that of San Diego, California. Thus, employees are asked to sign nondisclosure agreements and, when they change jobs, are "debriefed" as to what technology, if any, they may take with them. And the barriers are going up. Great efforts are going into the construction of "fire walls" between companies lest their employees drop secrets to the employees of a competitor. Major billion dollar corporations, such as Apple, hire private detectives to protect their proprietary ideas despite the negative press such methods evoke.

In the Technology War, industrial espionage has become as important as military intelligence is in conventional war. It becomes important for managers to have skills and sensitivities to these issues as well. Indeed, Admiral Bobby R. Inman (U.S. Navy, ret.), the first CEO of one of the earliest U.S. research consortia, was selected in part for his appreciation of the technology and in part for his extensive experience in military and international intelligence activities.

At the national level, what is needed is an understanding that each of the issues is important. The United States, with its reluctance to provide direct support to American vendors in international competition, its lack of coordination and decentralization, is seen as a Paper Tiger, whereas its adversaries, such as France and Japan, pick and choose their targets with coalitions of government and industry. The leaders of the French government helped persuade foreign buyers of the virtues of the Airbus, while American manufacturers received little support from their government. Yet the American aircraft industry contributes in a significant positive way to U.S. balance of payments. The U.S. government needs to play a greater role in helping its industries.

The Japanese use "administrative guidance" to segment their markets without compromising competition and use import controls to protect their fledgling and growing industries. Alternatively, the United States protects old and dying industries such as textiles and automobiles and places the high technology industries' products on the restricted lists, which impair their marketability.

Examples of more deliberate components of industrial policy, and their interaction with foreign trade and the domestic economy are discussed in the next chapter.

5

Industrial Policies

Industrial policy is a common but poorly understood term. To help alleviate this problem, we will define *industrial policy* as a set of interrelated policies that are intended to manage the development of a nation's economy, specifically its important industries and technologies, foreign trade, and domestic resources such as education and social programs in order to enhance a nation's well being. The difficulty then lies, first, in defining a desired level of well being and, second, in engaging in socially acceptable practices that will allow us to attain that level.

Adopting an industrial policy is a demanding task. The use of simple economic models will not work. Free economic systems deviate from the theoretical as we depart from the smallest economic unit such as a one-man business. Regulated and controlled domestic competition is obviously far from the theoretical level; international competition does not come close. Indeed, most industrial policies evolve from a set of disjoint government practices and laws.

Governments and the social and cultural systems that they are designed to protect and enhance play a critical role in a nation's industrial policy, especially in the manner in which major policies are established for the development of technologies. The governments manage their priorities through the use of incentives and counterincentives and actually have great leverage in determining how well one socioeconomic system will compete against another. Tax policies, antitrust laws, investment philosophies, and so forth, all play a crucial role in the rate at which any industry will develop.

MITI and Industrial Policy

Japan's industrial policy is shaped by MITI, which provides an institutional presence that holds many research and industrial programs together. (There is no comparable government organ in any of the Western countries.) The evolution of MITI and its role in industrial policy in Japan is especially important as well as interesting; for example, to some extent, MITI is an American product.

The roots of MITI's power can be traced back to the period before World War II. Japan, in the 1920s and 1930s, had an unusually controlled economy. Desperate to deal with their lack of natural resources, the Japanese had vested tremendous authority in the forerunner to MITI, the Ministry of Commerce and Industry (MCI).[38] MCI was born in 1925 and it ultimately evolved into MITI, one of Japan's most powerful agencies today.[39] Indeed, only Finance and Foreign Affairs match MITI in importance today.[7] As a measure of its importance, MITI employs a large number of graduates from the University of Tokyo Law School, second only to the Ministry of Finance.[39]

After World War II, MITI was under the direct control of the occupying authorities commanded by General MacArthur. SCAP, a name for both the Supreme Command for the Allied Powers and the commander himself, promulgated a variety of United States-like laws, including an antimonopoly law that was fiercely opposed by MITI. But perhaps no other SCAP action deserves more attention here than the Foreign Exchange and Foreign Trade Control Law that transferred controls from SCAP to the government through MITI. This enactment, in 1949, vested in MITI fundamental controls over foreign exchange and foreign trade transactions. SCAP had assumed that these controls could be reduced over time and had no idea they might "lead to the institutionalization of the most restrictive foreign trade and foreign exchange control system ever devised by a major free nation."[39] As a result of the law, Japanese bureaucrats in MITI were able to exercise far wider controls over the international relationships of their economy than any controls that could be exercised by officials in the United States.[7] Indeed, before the late 1960s, all technology that was transferred into Japan entered with MITI's approval on terms and timing, and all joint ventures that were approved and foreign patent rights that were acquired, were accomplished with MITI's forceful hand in the negotiations for some Japanese advantage. As a result, Japan acquired the rights to manufacture a host of high technology products with its (then) low-cost labor and manufacturing facilities. This set the stage for Japan's early successes in the international competition of the Technology War.

For many years, MITI used its informal rules to limit foreign ownership in Japanese corporations and foreign participation on the boards of directors of Japanese companies. Such control was compromised, however, when IBM wanted access to the Japanese market and Japan wanted access to IBM technology. In a working compromise, IBM Japan was capitalized as a yen-based company; IBM was then considered as a domestic organization. However, MITI was able to extract licenses to use IBM patents with very favorable royalty terms.[39] Apparently, IBM also complied with MITI's administrative guidance in exchange for access to the Japanese marketplace.

Administrative guidance is not a special power vested in MITI. It is more the embodiment of a tool manipulated by MITI to accomplish its objectives. Chalmers Johnson, an American authority on MITI, suggests that administrative guidance is really no more than the discretionary authority one entrusts to a diplomat. Success, he says, depends on the skill, good sense, and integrity of the diplomat.[39] It cannot be used too forcefully for fear it will provoke an overreaction that would undermine MITI's authority;

nor can it be effective without the responsiveness of the Japanese people.[7] Problems can arise if MITI's neutrality is suspect or when it is deemed to be functioning in support of another part of the Japanese government.

Many Americans tend to blame MITI for the "evils" committed by the Japanese in their so-called trading wars against the West. Considering MITI's scope, it is not surprising that it draws such fire. Its Heavy Industries Bureau, for example, has responsibilities for steel, machine tools, general machinery, automobiles, electronics, heavy electric equipment, rolling stock, and aviation products.

In fact, one of the most vitriolic attacks mounted against Japan and MITI was over textiles in 1969. Complaints came from as far away as the Union of German Textile Manufacturers.[39] Indeed, the terms "Notorious MITI," the "Ministry of One-Way Trade," and "Japan, Inc." were coined early in the 1970s, before semiconductors had achieved prominence in the marketplace. Still, MITI acts regularly to protect Japan's domestic industries. For example, despite a government announcement in 1973 that Japan was "100% liberalized" and, therefore, presumably open to trade, it still protected twenty-two industries of which seventeen (including computers) were the new strategic industries.[39]

U.S. Industrial Policy

The United States has no ministry that corresponds to MITI; however, it has been argued that

> the real equivalent of ... [MITI] in the United States is not the Department of Commerce but the Department of Defense, which by its very nature ... shares MITI's strategic goal oriented outlook.[39]

We agree that the Department of Defense (DoD) does plan and has played, a dramatic role in the development of information technology, but it hardly provides the national unifying force supplied by MITI in Japan. Nor does DoD get the respect and credibility that MITI commands by virtue of its role in the Japanese system. There is certainly no comparison in the U.S. academic communities where the Defense Department is considered a distant ally at best, whereas in Japan MITI is considered a good friend and close partner.

The fact is that the United States lacks a coherent industrial policy. Lee Iacocca, for example, wrote

> America already has an industrial policy ... or more accurately, hundreds of industrial policies. The only problem is that they're all over the lot and do little if anything for our basic industries.[40]

Industrial policy in the United States is exercised through its laws, departments of the government, regulatory agencies, and federal bureaucracies. Antitrust laws, for example, regulate commerce and competition in a decidedly Western way. In Japan, the industrial and banking federations known as *zaibatsu* institutionalize the Japanese mechanism for industrial collaboration under the watchful eye of MITI.

Antitrust law penalties and enforcement in the United States have been toned down recently to improve the competitiveness of U.S. firms. The abandonment of the largest legal case in the history of information technology, United States versus IBM, was indicative of one change; the recent wave of dumping and predatory pricing charges against the Japanese mounted by the International Trade Commission and the Department of Commerce are other examples of the government's desire to improve the competitiveness of U.S. firms.

Antitrust law is intended to facilitate market entry, competition through efficient production, and the establishment of prices commensurate with quality. Competition presumably invites technological innovation, and the antitrust laws have historically encouraged and protected competition in research and development. One disadvantage though, attendant with that protection, has been the discouragement of joint research ventures because they presumably might lead to collusion in the marketplace. This perspective is changing. In 1983, William F. Baxter, the Assistant Attorney General for Antitrust in the United States Department of Justice reported the following:

> the attitude in Washington is changing ... competition and cooperation are not ... necessarily inconsistent ... The creation and development of technology is one very important area in which such cooperation frequently may be beneficial ... so long as the sale ... does not unduly restrict the competition among alternative technologies or otherwise facilitate collusion among competitors ... the most efficient and effective means to develop new technologies should not be impeded.[41]

In another paper, Baxter outlined guidelines that the Department of Justice could use to determine if research consortia were anticompetitive.[42] Parameters such as marketshare and number of competitors could be used, he suggested, to determine, before the formation of a research consortium, whether such formation would invoke the ire of the Department of Justice. Since then, we have witnessed the formation in the United States of several major research consortia, such as MCC, MCNC, and the Software Productivity Consortium and the adoption of the National Productivity and Innovation Act. This act, among other things, removes punitive (triple) damages in antitrust issues related to the licensing of technology that is manifested as intellectual property.

Capital Practices

A nation's industrial policy, and thus its place in the Technology War, is also influenced by the direct and fundamental impact government has on information technology through its capital practices. These practices include the control of the currency,

exchange rates, savings rates, and management of the government agencies that loan and borrow money. Tax policy, federal
procurement regulations, SEC reporting requirements — even
environmental impact statements — can influence the use and
development of information technology. Furthermore, the relationship between defense and the domestic economy is important,
with the two sometimes inseparable.

Prior to World War II, a rich country and strong military
were synonymous in Japan. After the war, however, Japan was
forced to look to international trade to provide the wealth and
security for which it had previously relied on its military. Japan's
lack of natural resources, and the evolution of world markets into
high technology markets further drove the Japanese first into low
technology manufacturing, such as steel, and subsequently into
the high technologies. Their educated workforce, of course, was
ideally suited for such a migration. It is curious that the British,
who also have an island-based economy, did not respond to the
high technology challenge with the same fervor as the Japanese.

One reason may be the structure of the Japanese financial
system, with its emphasis on cheaper capital, which facilitated
the Japanese entry into high technology markets. It permitted
the Japanese to make heavy investments in production with relatively low cost capital.[43] Some of this cheaper capital came from
the well-known Japanese proclivity for saving.[7,14] The Japanese,
for example, work almost six days a week and save about 16 percent of their salaries, as compared to the Americans who work
less and save one-third that amount.[44] Bergamini offers Japan's
ability to sustain a spirit of cooperation for the benefit of the nation, despite being fiercely competitive, and the Japanese workers' willingness to sacrifice consumption for fringe benefits and
savings, as the main reasons that the Japanese growth rate of the
1980s can't be matched.[38] Some of their incentive to save may
come from a government policy that allows a Japanese family to
exclude from their taxable income the interest from over $60,000
in savings.[45]

Governments can do little to increase the total available capital but much to reduce its effectiveness. For example, since federal loans automatically create an advantage for one sector of the economy and a disadvantage for another sector, what is important is that the total available credit pool is rationally and systematically coordinated to meet national goals and that those goals are understood, well articulated, and supported by the people.[46]

Although we will not develop here an economic model of the capital forces that influence industrial policies, it is useful to examine them briefly. For example, price usually regulates imports and exports, which, in turn, control employment (at least in the manufacturing industries). Price depends on wages, taxes, and the relative values of currencies.

The manner in which the government manages exchange rates is an important component on industrial policy. The dollar may be worth 140 yen in 1987; in the late nineteenth century, Japan created its modern banking system with the yen valued at roughly half a U.S. dollar.[7] However, this high exchange rate was not favorable to a brisk export trade or the development of internal industrial growth. By driving the rate down (through the import of technology and industrial goods), the Japanese were able to get the added advantages of low prices, worldwide demand, and high employment.

It is important to note that Japanese are not always so successful; the reverse effect can be predicted. Japanese unemployment reached a twenty-five year high of almost 3 percent in 1986 because of the yen's 30 percent appreciation against the U.S. dollar in six months. In that period, American factory workers' average wages changed from more than 40 percent higher than Japan to 12 percent lower (measured in U.S. dollars).

Technology *should* reduce costs and increase quality in addition to establishing new markets and products. But technology is considered by some to be a threat to employment. Although it contributes to exports in one form, it can and does come back in another form in the guise of the competitors' products. Frequently, the returning product swamps the original product in terms of balance of trade. For example, $1 million of commodity products, such as microelectronic chips, can return as $1 billion in television sets.

All technical industries are vast consumers of capital; the capital investment required per worker has increased over 100-fold since the turn of the century, from $1,000 to $100,000 per worker.[6] And information technology is an even greater consumer of capital. For example, outfitting 300,000 new programmer positions with $50,000 workstations and networks (or capital equivalent facilities) would create a demand for about $150 billion — independent of the continuing demands for improved productivity for the existing workforce or for the required investments in improved computational devices, such as VLSI, architecture, and research. This is an awesome investment requirement to be placed on private enterprise.

To summarize, capital policies directly influence a country's position in the Technology War. Difference in exchange rates, salaries, and capital investments determine, to some extent, the relative efficiencies of the competitors. Equally important in the Technology War is the role played by foreign trade, especially now that quality and technology are so interrelated.

Trade Practices

Foreign trade is another mechanism that can be considered a component of industrial policy. Indeed, in the United States, foreign trade has traditionally been one of the most regulated aspects of the economy, and information technology has always received considerable attention because of its contribution to the trade balance.

At one time, American exports of computer hardware totally dominated the imports of computer hardware and served as a substantial contributor to the balance of trade. For example, in 1970, exports were approximately $1 billion whereas imports were negligible.[34] The ratio of exports to imports changed rapidly, however, swinging over 9,000 percent to almost parity in less than thirteen years. The trend is a national catastrophe for America. As it continues, we can expect continued calls for protection in American high technology trade with Japan and other Asian countries, especially Korea.

Controlling the Balance of Trade

One method for dealing with a balance of trade strain is to establish explicit trade barriers, such as import duties, quotas, and other restrictions. Another way in which to reestablish a favorable balance of trade is for a nation to reduce the value of the currency, such as Great Britain did in the 1960s and as the United States tried in 1985 and 1986. The exchange rate variations can have large effects on international marketing strategies. It usually takes twelve to eighteen months for their effects to propagate through the economy, however, in high demand markets it can take less time.

For example, in 1986, the yen experienced considerable appreciation against the dollar and became relatively much stronger in the United States than in Europe. This created, at least temporarily, a more favorable trade climate for the Japanese in Europe. In just four months, and despite a 20 percent duty imposed by the EEC as a defensive tactic, the Japanese quadrupled their sales of compact disk players. Such success was drawing dumping charges in less than six months from most of the industrial concerns of Western Europe. This response demonstrates that the commercial sector is much more sensitive to forces affecting their profits than the governments are that manipulate those forces. Nevertheless, in this case, the European governments obviously have responded to pressure from their industries. Curiously, the U.S. economy did not benefit from the devaluation that created this disruption in Europe so quickly, suggesting there are other reasons besides price in the competitive equation.

Another method for dealing with balance of trade problems is to erect import barriers that reduce the incoming flow of foreign goods. These barriers, however, are rarely sustainable over long periods of time; they usually break down, are circumvented, or result in repressive countermeasures. The heating up of the old American – Japanese commercial rivalries, the blatant French threats to veto Britain's imports of New Zealand butter, France's hostile treatment of Japanese video cassette recorder imports, and the exorbitant EEC tariffs on foodstuffs are all examples of crumbling policies and an imperfect economic model.

Of course, short-term effects of trade barriers can be huge and offer opportunities for arbitrage. When a currency fluctuates 20 percent relative to others in a few months, it is not uncommon for large profits to be realized. Sometimes these fluctuations are even engineered by government officials. In 1931, a Japanese group, led by government ministers, was anxious to recoup the losses they incurred when Great Britain abandoned the gold standard. Dismayed that they had not been warned in advance by the British, the Japanese concocted a scheme to sell yen secretly before the gold standard was abandoned by Japan. By purchasing foreign currencies and commodities with an artificially inflated yen, and then subsequently selling the commodities after devaluation, the Japanese conspirators realized a cash profit of over $200 million.[38]

As a rule, currency devaluations are expected to improve the volume of exports while depressing imports. However, the supply and demand experience in the market and government policy can also have impact on the results. It makes little sense, for example, to devalue your currency to encourage the export of high technology products, which, for defense reasons, you wish to discourage. Similarly, products of great commercial demand are likely to be sold regardless of the constraints placed on them.

In a non-Robinson Crusoe economy, it becomes very difficult to control the desired level of exports and imports. For example, the plunge in the value of the dollar in 1986 and 1987 made it much cheaper for U.S. trading partners to purchase American products. This advantage should have gone a long way toward helping the U.S. balance of trade in the short term. But it didn't! The Japanese have not responded to the cheaper dollar, and there

is increasing evidence that they simply are not interested in the
bulk of American products.

The Role of Quality

We have to inquire why the Japanese are not interested in U.S.
products. Quality is certainly a contributor to the decision equation, with the United States getting generally poor grades in this
category. Price is another. But the Japanese do not appear reluctant to purchase expensive goods. Rolex watches sell for four million yen ($30,000) in downtown Tokyo, despite the fact they can
be purchased for one-third that price at the duty free airports.
Expensive French brandies are the rage in the Ginza nightclub
district of Tokyo, and the fashionable department stores such as
Wako abound with luxury items from Europe. The explanation
is more cultural and is based on the nature of U.S. arrogance
which holds that the American way is the only way. U.S. products are simply not tailored to the special needs of the Japanese
markets. The Japanese just prefer to buy prestige products with
a reputation for quality. Most American products have an in-
flexible demand curve in Japan. They won't sell in the Japanese
market at any price!

On the other hand, the undervalued dollar is leading to major Japanese purchases of property and companies. The Japanese
are buying America, not American. Some of these purchases are
with an eye toward the future penetration of American markets.
For example, Fujitsu attempted to purchase Fairchild, a microelectronics firm owned by the French. It was only after the U.S.
secretaries of defense and commerce intervened that the transac-
tion was squashed in a widely publicized case that demonstrated
that American fears of foreign ownership in high technology industries does not extend to the French.

Even if the dollar–yen relationship produced the desired improvement in the American balance of trade with Japan, the
long-term prospects would be poor. The Japanese are not likely
to purchase finished goods. They are more likely to purchase

American high technology components that are not otherwise available in Japan, and the results of U.S. research which they clearly desire. Thus, the prospects for an improvement are dismal for America since it is likely that the exploitation of American technology will contribute ultimately to a greater marketshare for our competitors. Our technology will return in consumer products. Jobs will be lost and the deadly spiral will be once more unleashed.

Dumping

One of the more sensationalized skirmishes in the Technology War has been over charges of Japanese dumping of microelectronic chips into the American market. Numerous charges and complaints have been filed with the International Trade Commission and the Department of Commerce. The charges have alleged that Japan has sold chips below cost. There are several possible reasons for such a tactic, including gaining marketshare, maintaining employment, and reducing unit manufacturing costs by using economies of scale that result from increased production.

Dumping charges are very difficult to substantiate. How, for example, do we demonstrate that the Japanese are not properly reflecting their costs when we cannot even identify the real costs? One source in the Department of Commerce told us that there was particular doubt that the Japanese were marking up their prices sufficiently to reflect their research and development costs (similar allegations were made in the past with other technologies).

Some argue that dumping is a short-term problem that will ultimately bankrupt the dumper. How, it is asked, can the dumper continue to sell at below-cost prices? Others suggest that long-term marketing strategies that are based on selling below costs will continue only until the marketplace has been captured and the prices can be raised to generate monopoly profits.

An analysis of which condition might apply to a consumer in the United States is difficult. For example, consider another market in which the Japanese were also accused of dumping — the color television marketplace. At this time they certainly dominate this market. But, because of the fierce competition among Japanese manufacturers, we cannot tell if they are extracting monopoly profits and gouging the consumer or not. We wonder if the current low price for Korean television sets reflects their real costs.

As another example, in 1986, the EEC was convinced that the Japanese were dumping photocopiers in the European market. They imposed a historic 16 percent antidumping duty on twelve Japanese companies, arguing that the Japanese were selling 7–69 percent below their domestic prices. The Japanese response to this may be even more threatening to the European manufacturers' marketshares; Matsushita announced they were beginning the manufacture of photocopiers in West Germany in 1987.

Returning to microelectronics, the Japanese domestic marketplace is the second largest in the world. Yet, American products are not purchased in that marketplace. This may or may not be a quality problem, but, regardless, the United States has a special difficulty dealing with the Japanese: The only semiconductor suppliers that have been successful in the Japanese market are those that have established a permanent and significant presence in Japan.

The Japanese government claims that it can do little to influence Japanese buying patterns. Arguments are advanced that, besides inferiority, most Western products have not been properly adapted to the special needs of the Japanese markets. For example, some semiconductors are packaged with specifications written only in English.

The realities are that penetrating the Japanese markets is an extremely difficult task. We must not only have the quality image, we must penetrate the interlocking relationships among suppliers and vendors and one cannot threaten existing jobs. Further, we must have something they need. Mandating a market share, as was attempted between the Unites States and Japan in telecommunications and microelectronics, will not work.

The United States and Japan reached an unworkable settlement in 1986. The Japanese, without admitting that dumping of semiconductors had occurred, agreed nevertheless, to stop such practices and guaranteed the United States a minimum semiconductor market penetration in exchange for the United States dropping its dumping charges. But prices to the American consumer rose dramatically. In fact, prices for 256K RAMs increased 400 percent in three weeks after the settlement. The effect was to drive more American end product manufacturers offshore.

As this book went to press, the U.S. had imposed retaliatory tariffs on a range of Japanese consumer electronics products. This was expected to increase the prices to American consumers of some products up to 100 percent. We expect the Japanese to retaliate in an escalation of the Technology War. Unfortunately these acts and reprisals offer little hope for success unless they provide redress in two critical areas: opening the Japanese markets and increasing the competitiveness of U.S. companies.

The Japanese Miracle

Our analysis of Japanese and American practices highlights the role of society and government in the development and use of technology. There are also other arguments that explain the remarkable Japanese economic achievements. Four particularly interesting explanations that are labeled by Johnson in *MITI and the Japanese Miracle* are (1) the national character – basic values consensus, (2) the no-miracle occurred explanation, (3) the unique-structural-features reason, and (4) the free ride, at the expense of their allies.[39]

The first category, national character – basic values consensus, refers to the Japanese cultural capacity to cooperate and reach consensus on important issues. However, the consensus is, and has been, simply engineered by the government. *Bushido,* a form of Japanese chivalry that spread throughout most elements of Japan in the past, holds that the interests of the family and its members are inseparable. It is, therefore, easy to understand the natural sociological forces that the Japanese government could exploit to control and encourage consensus. In other words, there is a natural tendency encouraged by the educational and cultural system for the Japanese to cooperate for the good of society. In their exploitation, the government has merely recognized this tendency and geared itself up to facilitate that cooperation.

The second explanation, no miracle occurred, is advanced by economists who argue that the tremendous growth is merely the result of conventional market forces and the emphasis the government has placed on strategic industries. It may not be a miracle, they argue, and point out that it has happened before. For example, between 1850 and 1880, Japan reacted more quickly and successfully to Western ideas, economic principles, and military technology than any other nation in the world. By 1905 they were beating the Russians, and their navy was a match for most world powers, causing considerable consternation among many governments.[7] In fact, by 1919, the U.S. Senate was debating the threat to the United States of Japan's global adventures (Mexico, Panama, the Pacific islands, Siberia, and so forth), viewing them as particularly antagonistic.[47] Western resistance to Japanese expansion was actually being shaped early in this century.

The third explanation, unique structural features, is related to the first because it is based on cultural terms. Industrialists offer this argument focused on the "three sacred treasures" of Japan: (1) lifetime employment, (2) seniority wage practices, and (3) enterprise unionism, where unions participate in the management of the companies and permit workers to move from one job to another. These features also represent practices encouraged by the government.

The last category — free ride — is characterized by the opinion that Japan has benefited unreasonably from its alliance with the West and the relatively free American defense shield. This is not a new charge. Even Winston Churchill complained, in 1953, that the cost of armaments was interfering with the development of the British engineering industry, a disadvantage, he noted, that the Japanese did not have.[48]

Supporting this argument are some convincing data: U.S. defense expenditure per capita in fiscal year 1982 was almost ten times that of Japan and the ratio of defense expenditure to GNP was six times greater percentagewise.[49] Indeed, Japan refuses to commit more than 1 percent of its GNP to defense.

On the other hand, although less in dollar value, the Japanese are committing the same percentage of their GNP to research and development as is the United States. But when American defense research spending is removed from the calculations, the United States actually lags the Japanese in research and development spending. In fact, in the last twenty years, the American research expenditures as a percentage of GNP have remained flat, whereas the Japanese have doubled their expenditures. This trend is even more alarming if we consider its rate of change, which shows a significant growth in research expenditures for the Japanese but no change for the United States.

The open trading system in the West as well as the relatively open transfer of technology are also cited as additional examples of the free ride — that is, opportunities for Japanese exploitation. It is curious that the world seems to react so vehemently to these practices. These practices are not new! They are precisely those that the Japanese used in 1872 — sending students to the West and hiring Western experts.

Technology transfer has long been a major focus of Japan's industrial policy in this century and MITI makes no bones about it. They have stated their position with respect to technology transfer quite clearly

> [The] spirit of basing national development on technology should be our aim in the 1980s. Possession of her own technology will help Japan to maintain and develop her industries' international superiority and to form a foundation for the long term development of the economy and society.[39]

There is no Japanese miracle, and these are not idle words. The words have been matched by action! The Japanese have initiated a long-term strategy to win the Technology War, and they are using their industrial policies, social structure, and national infrastructure as part of that strategy. Western nations must understand that their historical dominance of world trade and technology cannot be automatically extended by reliance on the policies of the past. Technology has altered the strategic balance, like the long bow at Crécy which stunned the French and offered a tremendous short-term advantage to the British. Japan now has the advantage and Western nations must learn to respond with an integrated package of diplomacy, technology, strategic alliances, and financial and industrial policies.

That all of the issues discussed in this chapter are interrelated is not surprising; what is surprising is only that the governments of the West seem to treat each issue on an individual basis.

Summary

Since the Renaissance, the Europeans have stirred the world with their intellectual achievements. They have competed with each other in art, science, religion, philosophy, and economics and the results of this competition have been spectacular. The results dominated Western thought for five centuries.

Nevertheless, great intellectual accomplishment does not automatically result in the domination of world markets. This is further the case where technology demands the coordination of competition and a large standardized marketplace to provide manufacturing economies of scale.

The Technology War has confused the Europeans, and their constant intra-European skirmishes have exhausted them and fragmented their marketplace. The French, one of the more powerful players, bounce back and forth between the nationalization and liberalization of their economies and the rest of Europe is drowning in the government-imposed high taxes of socialism — used to subsidize a work force with out-of-date skills. A decaying industrial plant is milked for additional profits. In the global competition of the Technology War, the Europeans are a second-class power.

The Americans, on the other hand, have the advantage of an integrated economy, a large domestic standard marketplace, a superb scientific establishment, and great natural wealth. They should be in the choice position to dominate the Technology War.

Enemy action is not necessarily the only way to lose a war. One can also shoot oneself in the foot with an irrational national strategic policy. America has been doing this for some time through its rape of the nation's wealth in Vietnam and a welfare system that fails to motivate people.

Through the fog of all of this western confusion, the clear light of the Asian beacon illuminates our competitors who are focusing their resources, husbanding their weak industries, exploiting a work ethic unheard of in the West, and driving for technological supremacy.

The prospects of changing this dismal picture overnight are poor. Indeed, if there are any prospects at all, they must lie in education, the last hill in the terrain of the Technology War. But, as we illustrate in the next chapter, the educational systems in the West have alarming deficiencies.

6

Education

A nation's educational system is a critical element in its ability to function in today's high technology world and, consequently, to achieve and maintain a lead in aspects of the on-going Technology War. Advanced industries have special needs for trained personnel at all levels of the work force. There is, for example, a significant need for engineers and technicians; new graduates are continually needed to do design work, create manufacturing facilities, and maintain sophisticated systems; computer scientists are needed to invent algorithms, develop system software, write applications software, and document systems. In addition, as manufacturing systems are called upon to meet new tolerances and engineering constraints, people need to have more formal technical training. Factory workers who were trained in the 1940s, 1950s, and 1960s need to update their skills to meet the needs of the 1980s. Also, the use of technology is creating new job specialties for allied professions. Indeed, as high technology penetrates the office place and other conventional places of employment, new jobs, such as training workers, maintaining equipment, and so forth will also be created. For example, legal issues in computing have become a new specialty for lawyers, and, in fact, there are lawyers who spent most of their legal careers working on the recent United States versus IBM suit.[2]

American Practices in Education

The results of the U.S. educational system in science and technology in the twentieth century are impressive. Not only has the United States had more Nobel Prize winners than any other country, their achievements have been accompanied by wonderful technological innovations. Some of the notable inventions include the telephone, the light bulb, the moving picture, the airplane, and xerography. Names like Bell, Edison, Eastman, Wright, and Land are as familiar to the layman as their inventions.

Unfortunately, over the last fifteen years, there have been signs of a serious deterioration in technological innovation in the United States. Nobel Prizes seem to be more an accomplishment of the past than of the present. In fact, many of the recent prizes have been won by foreign-trained scientists.

Another indication of the decline in creativity in the United States is the number of patent applications. In 1985, 43 percent of all patents issued in the United States went to foreigners![50] This is a sad reflection on American innovation. Much of this deterioration in the excellence of American research is traceable to educational shortcomings.

There are several studies, such as the Gardner report, *A Nation at Risk*, that document the failure of the American secondary school system.[51,52] Some argue that the problems begin even sooner at the elementary school levels. In fact, the United States secretary of education has called for a major reform of curriculum citing a host of criticisms.

Only 16 percent of American secondary school students, for example, take any science or mathematics beyond tenth grade. And fewer and fewer American high school students elect mathematics and science courses. When they do, they are more likely to elect the life sciences rather than physics and chemistry. Only 33 percent choose chemistry and only 10 percent take physics. Fully 75 percent of American high school graduates do *not* have the prerequisites to enter a college science program.

When making comparisons with other countries, the U.S. position continues to slip. Japanese high school students take more advanced mathematics than most U.S. college students. France, Germany, and the Soviet Union require four years of biology, chemistry, and physics in high school. As for foreign language studies, the American efforts are overwhelmed by our aggressive and linguistically curious partners and adversaries.

A major problem is the quality of teachers in the United States. Not only has the public too little respect for teachers, many of our teachers are not good enough. What is cause and what is effect? It is hard to say, but it is clear that teacher prestige is high in countries that have good educational systems. The elementary and secondary school system in the United States needs major revision and perhaps even minimum national requirements for students who enter and graduate from high school.

Scientific and Engineering Education

There is overwhelming evidence that American scientific and, especially, engineering education is in serious trouble.[53,54] The United States once held a significant lead over all other Western nations in the proportion of scientists and engineers in its work force. In 1967, for example, the United States had more than twice the number of scientists and engineers engaged in research and development than the Japanese. But, by the early 1980s, the gap had shrunk to only 15 percent due to a decline in the United States and a rapid expansion in Japan.[50] Moreover, half the United States workforce was engaged in defense related work at the expense of contributions to the commercial community.

The problems with scientific and engineering education in the United States concern both quantity and quality. Therefore, the factors contributing to those problems also concern quantity and quality; they include the number of people graduating from U.S. schools with degrees in science, the number of foreign students in U.S. schools, the dropout rate of students, the decline in quality of both students and teachers, the deterioration of research equipment in schools, and the accreditation of schools.

In terms of producing college graduates, there are over 3,000 institutions of higher education in the United States. About 650 of them offer a master of science degree as their most advanced degree, and under 300 of them award the Ph.D. The doctoral-granting schools account for over 87 percent of all of the graduate enrollments.[55] Curiously, private institutions play a disproportionately large role in graduate programs in science and engineering in the United States.

In the early 1980s, the number of undergraduates in engineering departments in American universities began to increase rapidly. For example, in 1973, there were 187,000 full time engineering undergraduates, whereas in 1982 there were 403,000.[56] According to the 1986 New York Times Supplement on Education, computer science enrollments increased by almost 600 percent between 1973 and 1983. This meant that as many students enrolled in computer science as in English, the life sciences, the physical sciences, and psychology.

This might appear to be good news, but under scrutiny, some distressing factors come to light. For example, while the number of master's degrees in engineering increased slightly in the 1970s, the number of doctorates actually declined, contributing to a serious decrease in faculty in the engineering colleges. In addition, the proportion of Ph.D. degrees in engineering that were granted to foreign students increased. Between 1980 and 1983, there may have been a 6 percent increase in graduate enrollment in science and engineering, but the number of American citizens in this group was up only 1 percent. Foreign students represented 85 percent of the total growth in this category during this period.

The presence of foreign students in American engineering schools raise complicated questions. Do they represent a brain drain away from other nations such as Britain, India, and Third World countries? Do they fill a critical gap in the American workforce? Do American schools present attractive opportunities for foreign students who would otherwise be unable to contribute technically in their own societies? Harvey Brooks, a former dean at Harvard University, asks whether encouraging more American students in science and engineering would lower educational standards since, at least on paper, the foreign students have superior qualifications.[57] On the other hand, if we continue to train foreign nationals, are we training future competitors? Are we creating allies and future markets for U.S. products or strengthening our adversaries at our expense? We don't have the answers to all of these questions. Certainly, additional objective data and analysis about the career patterns of foreign and domestic engineering students would go a long way towards filling that vacuum.

Another factor contributing to the decline of technically educated people in the United States is the dropout rate. Some of the best American graduate students in computer science leave school before obtaining their degrees. One explanation for this behavior is the entrepreneurial adventure that attracts some of the most creative minds. An example is Steve Wozniak who dropped out of the University of California at Berkeley as an undergraduate to help found Apple Computer. The success of role models like Wozniak, as well as Andy Bechtolsheim and Bill Joy, dropouts who founded Sun Microsystems, has encouraged other talented students to leave school without their degrees.

There are other reasons for students dropping out of school, especially graduate school. Industry, for example, offers excellent salaries to people with undergraduate degrees. Doctoral degrees may be useful for teaching jobs, but the poor working environments in many universities offer little incentive to seek an academic career. Further, overworked faculty members are poor role models for students making career choices. Although a master's degree does give students an educational advantage in an industrial career, the salary advantage may not be sufficient to justify waiting an extra year before entering a company. This is particularly true for people going to work in new companies. Indeed, students with advanced computer science degrees can command salaries that are competitive with those of MBAs!

Although the quantity of students in areas like computer science is large, the quality may have declined. This fact is hard to document because on paper, today's graduate student looks better in terms of grades. But grades have been debased, partly as a result of a long process that began during the years of the Vietnam War. In those days, a student who flunked out of school might have been drafted and sent to Southeast Asia. Members of the faculty, not wanting such responsibility, were susceptible to the pleas of the form, "If I don't get an A in this course, I'll be forced to drop out of school, and go to Vietnam." The result was a rather obvious escalation of the grade point averages among the student populations.

The quality and quantity of faculty members in engineering schools are additional issues related to the problem of technical education in the United States. In an important study, F. Karl Willenbrock, dean of engineering at Southern Methodist University, stated, "The most crucial problem facing United States engineering schools is their decreasing ability to attract and retain faculty members who are in the top rank of their technical specialities."[58] Moreover, he inferred that faculty careers are no longer held in high regard by the most capable Americans. He argued persuasively against the trend of hiring non-United States citizens into engineering departments because of the difficulty this creates in maintaining the close ties between industry and the Department of Defense that have benefited colleges of engineering in the past.

Despite the explosive student growth in the nation's engineering schools after World War II, many engineering schools have had difficulty attracting and then retaining first-class faculty members. As a result, there are presently 2,000 unfilled faculty positions in engineering in the United States. Some of the reasons for this include poor working conditions in academia, low pay, excessive work loads, outdated equipment, and limited research funding.

These factors have also accounted for a significant reduction of quality in academic programs. As a result of the shortage of teachers, some departments, desperate to hire faculty, lowered their standards in the mid-1970s. This, in turn, led to the staffing of many departments with faculty who are adequate teachers but who lack research credentials. Furthermore, their presence has continued to discourage talented young graduates from taking academic jobs.

Another problem in U.S. institutions of education is organization and resistance to change. The engineering faculties, in particular, have a large number of people nearing retirement who, at that stage in their career, resist changes in organization and in the allocation of resources. Of all U.S. universities, only MIT and Stanford seem able to adapt to new technological areas.

On the positive side, there does appear to be a consensus now that reorganization of engineering schools is necessary, and the National Science Foundation is funding the development of innovative Engineering Research Centers.[54,59] Integrated centers such as these have been initiated in biotechnology, telecommunications, robotics, and advanced materials.

Other reasons for the shortage of technically advanced people in the United States include the decline in the amount and quality of research in academic institutions. One reason for this decline is the deterioration in the state of the art research equipment available to American universities, a serious disincentive facing a young Ph.D. contemplating an academic career. Research facilities have deteriorated since the last serious federal investment in university research equipment which occurred in the late 1950s and early 1960s. Then, during the Sputnik era. there was major federal investment in the research and development plant at colleges and universities. In 1963, for example, the U.S. government spent over $200 million on university and college research and development (R&D) plants. In the 1970s, however, the federal priorities shifted to other issues, such as crime, welfare, and energy, and by 1981 only $25 million was spent on R&D.[54] We are faced now with a major problem in the physical plants of our universities and a need to reinstrument our university laboratories. Over a billion dollars may be needed to solve this problem.

In human terms, a young graduate making the choice between taking a job at a university or at an industrial laboratory will quickly discover that the level of capitalization available per researcher in industry is often two to five times that in an academic department. It is no wonder that the best people in many computer science faculties choose to perform theoretical work because such research does not demand advanced computational equipment.

Another indication of the quality in American engineering educational programs is accreditation. Over the years, the percentage of departments accredited has held steady at 70 percent yet in 1981 only 50 percent of those departments reviewed were accredited.[24] This could reflect deteriorating performance or increasing standards. In either event, they are a poor reflection on U.S. schools.

Many engineering schools teach a brand of applied science that is often neither rigorous nor useful. This leads to the question of who should teach practical engineering? Industry wants students who have learned the latest applied skills. Faculty members argue that the role of the university is to teach fundamentals, not to give vocational training. This tension between pure and applied research and relations with engineering is discussed by Shapley and Roy, who argue for a total reorganization of the United States university science systems.[59]

What, then, are the future prospects for America's engineering schools? Unless the present situation, which has been stable for a number of years, changes, these schools will continue to atrophy. Willenbrock goes so far as to suggest that the system may evolve into one similar to the French approach, which has a small number of *grandes écoles* that are good teaching institutions.[58] The problem is in France, the *grandes écoles* are not research institutions. In contrast, in the United States the top schools are also the premier long-term research institutions. Thus, if we adopt the French system, the number of basic research institutions in America would decline.

It has also been suggested that we should encourage talented young students to study pure science as undergraduates and let industry provide the engineering education. One development along these lines was the establishment of the Wang Institute, an accredited school giving a master's degree in software engineering. The school was founded by An Wang of Wang Computers to specialize in an area that had been ignored by most traditional engineering colleges. Unfortunately, this novel experiment was prematurely terminated as a result of financial problems encountered by Wang.

Continuing Education

Since knowledge is expanding at a rapid rate in the high technology areas, continuing education programs have proliferated. No college education, no matter how good it is, can suffice for a forty year career in rapidly developing areas. Thus, technical people must renew their educational background periodically. To meet this need, many colleges and universities have extension programs for part-time undergraduate and graduate courses. These continuing education programs have all of the difficulties found in regular academic programs, which are further exacerbated by a part-time faculty, more diverse student body, low prestige and high instructor turnover.

Another source of continuing education are the scientific and professional societies, which provide short courses, tutorials, and professional development seminars. There are also seminars operated by individuals and companies on a for-profit basis, such as the James Martin Seminars. The quality of these seminars is often higher than the tutorials run by the professional societies because the lecturers are better paid.

Japanese Educational Practices

Japanese educational practices contribute significantly to Japan's success in the Technology War. For example, Japanese high schools provide a substantially better grounding in mathematics, statistics, and basic science than almost all American high schools. In fact, Japanese secondary school students consistently score higher on standardized mathematics and science tests than any other students worldwide.[8] Japan has a high degree of standardization in its high school curricula as well. Also, because of the stringent requirements of the university entrance examinations, there is a much smaller variation in Japanese students' backgrounds than in the United States.

In both the United States and Japan, institutions of higher education vary in terms of prestige. However, the status hierarchy of universities is particularly clear and rigid in Japan. The "big seven" schools are Tokyo, Kyoto, Osaka, Tohoku, Hokaido, Nagoya, and Kyushu. The University of Tokyo clearly leads the pack and is said to have as much prestige in Japan as Harvard, Princeton, and Yale *combined* would have in the United States.

In engineering, the top universities are the two oldest Japanese universities, Tokyo and Kyoto. By contrast with the United States, government-supported institutions have greater prestige than private universities; there are only three private schools among the top twelve engineering schools in Japan. As in the United States, private institutions charge much higher tuition fees than public schools. Thus, in Japan, one finds the schools of lesser quality and reputation charging more fees than the schools with greater prestige.

Faculty and Research

The role that faculty members play in Japan is strikingly different from that played by professors in the United States. Japanese professors, for example, have a very indirect influence over industry. There is no Japanese counterpart to the American faculty members who actively manage companies "on the side," or who consult extensively in industrial research. Japanese professor's principal influence over industry is essentially through their participation on government steering committees for MITI or by their students' contributions when they enter industry.

There is also a more pronounced difference between public and private universities in Japan and America. Faculty at the national universities are government employees, and they cannot accept paid consulting or advisory assignments for private companies. Nor can they easily accept major research grants from private firms. Of the expenditures on research at the national universities in Japan, over 98 percent comes from government sources. Such dependence on government support explains how the government can influence the role national universities play in meeting national objectives.

The private universities, on the other hand, are less restricted, and research funding from industry is extremely important to them. In these universities, only 16 percent of research funding comes from the government, whereas over 83 percent comes from industry and other private sources.[60] It is for these reasons that professors from the private schools are heavily involved in relations with industrial sponsors.

Prestige and Status

In the United States, the best universities are the private ones because there are fewer outside political and bureaucratic constraints. In Japan, it is just the opposite, the reasons being tradition and funding. It is prestigious to be a professor at a public university like Tokyo or Kyoto. In the United States, perhaps the only comparable public university is the University of California at Berkeley.

The status of the institution from which an engineer receives his bachelor's degree has much greater impact on a professional career in Japan than in the United States. First, the major companies hire their engineers directly out of the university. Second, they rely heavily on the status ranking of the university to judge the quality of the engineer. Still, there have been some new and interesting trends. Some companies now require examination of prospective employees. Such an idea has been occasionally proposed by some American companies to overcome the tendency of professors to write glowing references for all graduates.

Since the introduction of Western technology into Japan during the Meiji period, the engineering field has enjoyed relatively high status in Japanese universities and continues to attract large numbers of students. In fact, engineering is one of the paths into management in the large manufacturing companies of Japan, and therefore, it attracts potential managers as well as future engineers.

Graduate Degrees

Although Japan has a smaller proportion of its population going on to higher education, Japanese universities have been graduating more electrical engineers (in absolute numbers) than American schools since 1973 (See Table 6.1). They have led the United States in producing bachelor's degrees since 1970. They are, however, lagging in the production of master's degrees and are still considerably behind in the production of doctorate degrees.[60]

Table 6.1 United States–Japan Annual EE Graduates[60]

	United States			Japan		
Year	BS	MS	PHD	BS	MS	PHD
1969	11,375	4,049	858	11,035	703	108
1970	11,921	4,150	873	13,085	688	116
1971	12,145	4,359	899	14,361	844	109
1972	12,430	4,352	850	16,020	913	119
1973	11,844	4,151	820	16,205	1,026	114
1974	11,347	3,702	700	16,140	1,173	106
1975	10,277	3,587	673	16,662	1,258	120
1976	9,954	3,782	644	16,943	1,201	114
1977	9,837	3,674	574	17,868	1,447	142
1978	10,702	3,475	524	18,308	1,686	132
1979	12,213	3,335	545	19,572	1,697	166

Doctoral programs in Japan differ considerably from those in the United States. Japan adopted the European model for postgraduate education in the nineteenth century, and its influence lingers. For one thing, only students aiming at academic careers pursue full-time doctoral studies. In addition, Japanese doctoral programs consist not of formal courses but of the student serving as a research apprentice to a faculty member for an undetermined number of years. The student serves until he produces a piece of research of sufficient substance to win the doctorate. Doctorates are also conferred on researchers in government or industry laboratories who make major research contributions.

Chinese and European Practices

Despite long traditions of science and education in Europe and China, there are serious deficiencies in their educational systems where high technology is concerned. A rigid social system, a characteristic of Europe, affects some of Europe's educational practices in a deleterious way. For example, it is considered *declassé* to be an engineer in British society. And China is only now recovering from the excesses of the Cultural Revolution and its attack on independent thinking and education.

People's Republic of China

Because of extremely high unemployment rates among young people in China, which are even higher than the estimated 20 to 25 percent for the overall population, competition is stiff for entrance to universities. Only five out of every hundred applicants succeed in gaining admittance to Chinese universities, all of which are public. Unfortunately, due to a complex grading scheme and petition process for admission, even a very good student who has done well in his or her entrance exams might be rejected if the choice of university or the department in the university is oversubscribed.

Computer science is considered important in China, and upon entry the average student in computer science is very good. However, the education they receive is poor and does not meet Western or Japanese standards. For example, the average computer science student in a Chinese university gets only about forty hours of computer usage time during the entire four-year program of study whereas in Western or Japanese universities the students gets hundreds of hours. Further, nearly all the computers used by the students in Chinese universities operate in a batch mode with paper-tape input, a very old fashioned and not very productive use of those forty hours. Therefore, although the computer science curriculum in China is based upon the de facto Western standard (the Association for Computing Machinery model curriculum), the lack of exposure to computing facilities render the courses less effective for the student than they could be.[61,62] The Chinese must overcome this deficiency in their facilities if they hope to compete in the Technology War.

Another problem with the Chinese education system is the lack of incentives. Although it is true that entrance to a Chinese university computer science department is difficult, once a student has entered, the competition relaxes. Life in a university is quite good by Chinese standards. The student pays no tuition and gets partial support for room and board. Although the school year is long, forty-four weeks per year at six and one-half days per week, the demands placed upon students are minimal by American standards. Further, the students are guaranteed a job upon graduation; high scholastic achievement does not necessarily result in a better job or better paying job. Finally, the quality of this generation of teachers is poor; the Cultural Revolution effectively decimated the generation that today would be its prime teaching years.

Coupled with the students' lack of economic incentives to excel is an apparent lack of intellectual curiosity to learn more than the teacher offers in class. According to a Chinese computer science lecturer who recently returned to China after a two-year stay in the United States, the average Chinese student is spoon-fed detailed knowledge. The teacher prepares laborious notes for every lecture, and the student memorizes everything that is presented, but no more. Since there is no opportunity for creative work on a computer terminal, there is no such person as a "hacker" in Chinese universities. This is in sharp contrast to the students, faculty, and postdoctoral visitors who attend American universities from China. Their intensity, motivation, and desire to learn *everything* possible during their visits is astounding.

China has recently instituted postgraduate education in computer science in many Chinese universities. Most of these institutions will be offering master's degrees; only a few will be offering Ph.D.s (the first Ph.D. in computer science in China was awarded in 1982).

Will China surface as a contestant in the Technology War? It is hard to say, but China's long traditions of intellectualism are likely to prevail over current politics. Nevertheless, it is a long march from their present position to a competitive place in the Technology War. China runs the risk of the poker player drawing to an inferior hand; its cards may improve but the other players already have the advantage.

Great Britain

Great Britain's educational system should be outstanding: It has produced a disproportionate share of world class scientists based on the size of its population. Indeed, Britain has made fundamental contributions to computing, dating from Babbage, Lovelace, and Turing to the present day. The system is geared to outstanding training for the elite. However, its educational system is lacking in many respects: The supply of ordinary engineers, scientists, and technicians is inadequate, in part because such people fail to command esteem in British society. In fact, Great Britain graduates only 6,500 engineers in information technology annually. By way of comparison, see Table 6.1 for the number of graduates in the United States and Japan.

The British approach, which is very conservative in making

capital investments in general, is even more so in making educational investments. Indeed, in recent years, there has been a noticeable decline in the quality and quantity of British science. This conclusion has been supported by a government report to the secretary of state for education and science, which argues that cuts in British government spending have resulted in a real 5 percent reduction in research and development spending over the last six years. A Technology War is no time for a contraction of this magnitude. Great Britain's future as an industrial nation will not be ensured by such declining investments.

Most new British Ph.D.s often acquire research experience abroad, and the young scientists are reluctant to return home. One reason is that a new Ph.D. receives a starting salary of $12,500 in a British university and two or three times that in an American one. In fact, some American schools start assistant professors of computer science at over $40,000 per academic year. Furthermore, in the United States, some industrial starting salaries for Ph.D.s from top ranked schools approach $60,000 per year! These salaries are comparable to the starting salaries for graduates from some of the best business and law schools in the United States.

Salaries are only one facet of the comparative problems facing British professors. The young British professor at home must cope with outmoded equipment and a lack of long-term prospects for adequate research funding. At a time when the United States, Japan, and other countries are increasing their expenditures in research and development, Great Britain has been reducing its investment. The consequences of such a misguided policy decision are an inevitable decline in British science and technology. Unfortunately, the British government may have no choice, simply because of a lack of national resources. The respected Alvey report focused on these educational shortcomings in Great Britain, and funds have since been programmed for about fifty new professorships in British universities.[63] It remains to be seen whether this investment will make it possible for Britain to pull out of its educational tailspin. The prospects of that seem poor.

Federal Republic of Germany

West Germany has a solid elementary and secondary school program. Students entering the university are well grounded in mathematics and science. The German engineering tradition survives, and schools such as the Technical University in Munich and other technical universities as well as the universities of Berlin, Karlsruhe, Erlangen, and Saarbrücken are considered quite good in the international community.

Germany was quick to realize the importance of computer science (or *Informatik* as it is called in Germany) in the early 1970s. A national program was started in numerical analysis, mathematical physics, and parallel computation. Professorships were made available in many universities, and these very attractive lifetime positions were quickly filled with the best available people. Unfortunately, the candidate pool at that time was small and the quality of the appointees was below par. The consequence today is that the middle and senior ranks of German universities have no vacancies for attracting young and innovative scientists: Younger people have little incentive to join the universities since it could be many years before they can matriculate into senior positions in the German educational establishment.

The United States has benefited from the immigration of outstanding young German scientists because of the lack of opportunities in their country. It seems to be common for the Germans to export the cream of their scientific crop. Historically, United States science owes much to European emigrés from Germany who participated in the Manhattan Project, such as Szilard (who, with Einstein, advised Roosevelt to build the bomb), Teller (the father of the hydrogen bomb), Bethe, Franck, Nordheim, Rabinowitch, and so on.[27] Also, the United States (and Soviet) space programs included many German scientists that cut their teeth on the V1 and V2 programs of World War II.

Relations between academia and industry are rather distant in Germany. In private conversations, German professors criticize industry for being short sighted and disinterested in research.

On the other hand, German industrialists claim their academic counterparts to be impractical and uninterested in real problems. Still, American academics who visit Germany are impressed by the esteem in which professors are held there. German academic salaries are generous. It is considered a very important moral obligation of the federal government to pay the salary of its professors. Therefore, a German professor can obtain a bank loan to buy a house with no down payment. Such government and social support of the educational institution is an important asset in the Technology War.

France

France has a distinguished scientific history and a strong traditional educational system. French history in physics and mathematics is, perhaps, unparalleled, and French students who enter American universities are well grounded in these topics. However, the rigid French examination system stifles some creativity in students, as contrasted to the English system, which encourages it.

It is curious that France has a good and centralized educational system. In the United States, the traditional wisdom has always been that a centralized system would lead to a poorer school system. Perhaps the assumption is based on an assessment of the quality of the bureaucracy that would implement the national testing and standards that must accompany such a program.

Some outstanding French students have been educated in the better American schools. However, the American computer science Ph.D. degrees, even from the best schools, are not recognized by French academic departments, and those graduates who wish to pursue an academic career must obtain a French doctorate as well. This requirement is counterproductive since there are people with good credentials, ordinarily at the cutting edge of current research, who are forced to subordinate their work to the management of a French professor for several years. Many of the more talented people refuse to accept such demeaning working conditions and go to work in industry. This reflects poorly on an otherwise excellent educational system.*

One of France's major problems is the lack of an adequate number of engineers. Rigidity has impaired the flow of new professorial talent into the French engineering schools. Only recently, when France recognized that it lagged in the technology race, did it become concerned about its inadequate supply of engineering graduates. Compounding the problem, a demographic study of French engineers showed that a large number of France's older engineers were soon to retire, and there was an inadequate number of French replacements for them. To fill this impending need, one plan put forth by the government was to establish three additional engineering schools. However, there was opposition to this plan from the present engineering schools. It appears at this time that the new engineering schools will not be established; rather, the funds programmed for expansion will be spread around among existing schools in an attempt to increase their production of engineering graduates.

Another idea under consideration is to reassign some French national universities to the provincial governments. This plan has a financial advantage to the government since it permits discontinuation of their federal support of these schools. It is hoped that regional pride will motivate taxpayers in the provinces to pay for these additional schools. There is some precedent; a number of years ago, some of the highway network was turned over by the national government to the provinces. Provincial pride motivated the taxpayers to maintain and improve *their* roads. It remains to be seen whether the French will recognize that the educational infrastructure is as important in the Technology War as the transportation and manufacturing infrastructure is in conventional war.

---

*Notwithstanding this elitism, it is interesting to note that the University of Nantes was prepared to award a French Ph.D. for a doctoral thesis questioning the existence of Nazi gas chambers.[64] Only the last-minute intervention of the government prevented the granting of the degree.

Results of Educational Practices

Good ideas in basic science in the United States are most often developed in a university. The American and Western educational system produces a host of students and research results. Consequently, many American and Western companies rely on university sources for pure basic research. In Japan, on the other hand, much basic research is conducted in the companies, and the Japanese universities perform almost none. This is reflected in how engineering is taught as well.

In Japan, graduates exit from a university training program with broadly based and extensive training in theory but, with fewer specialized and applied courses than their American counterparts. This education provides flexibility that is more useful in the general careers that await most Japanese graduates. But it also means that employers need to provide graduates with more time and structured orientation before they can be expected to make an independent contribution. Furthermore, since the best engineers in Japan are in companies, practical techniques can be taught in an industrial setting by experienced design engineers. Since a Japanese employee rarely leaves the company, this investment in pragmatic engineering is unlikely to be wasted. The result is that the Japanese engineer can and does receive a better education in practical engineering than an American student.

Research ideas are developed in concert by the research community in Japan. But there is no stigma associated with engineering and, because of lifetime employment practices, people can and do stay with their ideas as the technology migrates from research through advanced development through production to the marketplace. This produces a significant advantage for the Japanese in technology transfer. Once past the research phase, it is plausible to inject new ideas anywhere in the technology transfer cycle in Japan. In addition, Japanese workers are notably less resistant than employees in other countries to the acceptance of new technology.[65] This simplifies the use of other people's ideas and provides good incentives to have an open mind.

There is less influence from lifetime employment practices in Europe than in Japan, but more so than in the United States. The European forces tending toward single employers in the past have included the lack of alternatives, stable social benefits programs (especially in Scandinavia), a lack of entrepreneurial opportunities (in particular, a lack of venture capital), and some language barriers. However, the forces are not as strong as in Japan, and the result is that Europe falls between the United States and Japan in lifetime employment practices. This leads to a more Western approach to industrial training in Europe. Unfortunately for Europe, the NIH syndrome, is made worse by the entrenched conservative scientific establishment.

Summary

Education is a key strategic factor in the Technology War. It is the secret to the supply of the troops necessary to fight the war on all of its fronts. The next generation of Americans will pay a terrible price if we ignore the problem today.

The United States has an inadequate elementary and secondary educational system, and its higher education system is of mixed quality. The best of the universities are the best in the world, but the quality, in computer science, for example, drops off abruptly after the top four ranked institutions. Engineering education is poor and needs innovation. The United States is too dependent on foreign students in engineering. The European educational systems are the greatest weapon in Europe's arsenal. Germany and France are strong in both science and engineering education, and Britain is strong in science. Japan has a good elementary and secondary school system. The Japanese universities are good but do a poor job in training Ph.D.s and researchers. The Japanese formal educational system does not adequately support its national aspirations in research but Japanese industry produces superb engineers. All the countries are weak in providing continuing education for engineers.

There *is* a direct link between educational quality, inadequate engineering graduates, uncompetitive products, marketshare, and unemployment. Most of our competitors recognize this and are adapting their educational systems to further their ambitions. This is especially so in places like Singapore, where they see obvious commercial advantage, and in Japan, where they are willing to wait generations to meet their objectives. The United States, too, needs to recognize the value of education in the Technology War and to revamp its educational system. Over the last twenty years, taxpayers have not been generous in their support of the public universities. They, too, need to change their attitude.

In the next part of this book, we introduce the various government and industrial combatants in the Technology War. These warriors operate in the terrain that has been described above: education, industrial policies, and the factors that shape technology transfer. These issues set the stage and need to be understood, but it is the current research efforts and the strengths and weaknesses of the great industrial powers that represent the front line combat.

PART III
THE WARRIORS

7

United States Government

Because the Technology War is so capital intensive and because it requires all the forces that a nation can muster to win it, the governments of the world are the major competitors, or warriors, in the war. In both Japan and the United States, major government agencies serve as the grand marshalls, whereas in Europe, in a modest attempt at coordination, the EEC is the grand marshall.

There are many organizations in the United States that can claim the title of the lead agency in this technology competition. The National Science Foundation, National Aeronautics and Space Administration (NASA), and the Department of Energy could all lay claim to some of the recognition and, indeed, are entitled to coverage in this chapter. But any account of the development of information technology in the United States must start with, and give most of the past and current credit, to the Department of Defense.

Department of Defense Programs

The Department of Defense carries the burden for protecting and defending the United States and its allies against adversaries. In both conventional and nuclear war, the major antagonist for over forty years has been the U.S.S.R. This situation has had a tremendous influence on the Department of Defense's approach to technology since, throughout this period, the Soviets have enjoyed a significant advantage over the United States in personnel, deployed military units, and conventional weapons. NATO as well has suffered a severe disadvantage in conventional arms in comparison to Soviet and Warsaw Pact forces. And the deficiency in armor is only one of many deficiencies in conventional forces that has compelled the Americans and their allies to seek "force multipliers" through the use of technology.

It has been concluded that only the use of high technology in advanced weapon systems offer the United States the potential for superiority that is needed in the field. According to Thomas Etzold, this explosion in the need, deployment, and potential use of modern high technology weapons

> may well be the most important military development since World War II. It exceeds, in some respects the significance of evolution in the nuclear forces of the superpowers, and it possesses higher consequences than ... nuclear proliferation.[66]

Thus, the Technology War is, to some extent, an evolution of conventional warfare, where Western countries have tried to fill the perceived gap with the Soviets with technology. It may be this same gap, viewed as a strategic advantage by the Soviets, that has led them to de-emphasize technology with the unforeseen consequences, perhaps, of having lost their place in the Technology War.

There are many that argue that the use of high technology only increases the costs of weapons and reduces their effectiveness because of the reliability and maintenance problems attendant to high technology systems. The Soviets, they say, have a much greater ability to concentrate firepower since their weapons, free of sophisticated technology, are more reliable and are not as likely to be down for repair as often as Western arms. This may be true in a conventional military sense and, if so, the West's only option is to improve the reliability and redundancy of its technology. As a consequence of this debate, these problems associated with technology, from development to reliability, have received considerable attention in the defense community. The result is a highly motivated and active defense research activity. And, within the Department of Defense research and engineering establishment is the Defense Advanced Research Projects Agency, the flagship research organization of the office of the secretary of defense.

The Defense Advanced Research Projects Agency

The Defense Advanced Research Projects Agency (DARPA) is one of the most important research funding agencies in the Department of Defense. It was created in 1958 to carry out long-term, high-risk, and high-impact research and development. The agency first concentrated on space-related technologies such as satellites and launch vehicles. However, after NASA was formed, many of the space-related research projects started by DARPA were assigned to NASA. DARPA then concentrated on such topics as ballistic missile defense, computer technology, and other high technology problems that could offset the Soviet advantage in conventional forces. It is important to note that DARPA's success in computer technology is one of the greatest accomplishments of any government-funded activity in any field, matched only, perhaps, by the Manhattan Bomb Project and the great work in epidemiology and research in communicable diseases by the National Institutes of Health.

DARPA manages basic research and advanced development in industry, at universities and other not-for-profit organizations, and at government laboratories. In 1984, roughly 75 percent of DARPA funding was spent in industry; the remainder was split fairly equally between universities and not-for-profit organizations; very little was directed at in-house government research. DARPA's emphasis on basic research is typical of the government in general; for example, in 1986, almost 64 percent of all government funded work was classified as basic research.[67] Consistent with that emphasis, the staff at DARPA has been composed of well trained and motivated personnel with technical and management skills generally superior to those of people found in other departments and agencies of the Department of Defense or, in fact, the rest of the government.

DARPA's accomplishments include the development of time-sharing, artificial intelligence, the concept of packet switching, and networking. One of DARPA's most successful projects was the development of computer networking. Packet switching, an efficient networking method that replaced conventional circuit switching for many computer applications, was explored, developed, and used to connect advanced research computers across the United States. The first network, built in the 1970s and known as ARPANET, was used for computer-to-computer communication. It became a powerful computer science research tool in its own right. Subsequently, ARPANET was expanded beyond the research community to include quasioperational users; part of it later became the communication backbone of the Defense Department's Digital Data Network. In the late 1980s, many derivatives of ARPANET were available throughout the world in commercial versions such as TELENET.

Strategic Computing Initiative. In 1983, two years after the Japanese startled the world at their Fifth Generation Computer Conference with their ambitious research plans, DARPA unveiled a ten-year plan called the Strategic Computing Initiative (SCI). The level of initial funding established for the program was $300 million for the first three years, a sizeable amount for an agency whose overall annual funding had been typically in the $1 billion range.

The SCI plan included using new material and fabrication techniques to build microelectronic chips, which are both fast and hardened to withstand radiation, and to develop parallel or concurrent architectures that would offer enhanced speeds. DARPA made it clear that basic research in expert systems and reasoning would be continued, along with work in machine vision, speech, and machine understanding of natural languages. However, DARPA, in contrast to its historical emphasis on more conceptual work, specified that the future results would have to be demonstrated in practical applications. DARPA's increased emphasis on applications and demonstrations was a reflection of a growing recognition in defense circles of the need to increase the flow of technology from the laboratory to weapon systems.

The SCI plan established three military applications goals to demonstrate the research objectives.[68] The first was an autonomous land vehicle that could be used in an environment such as a battlefield or some other area that might be exposed to high levels of radiation. Such systems must have the ability to plan and reason using data acquired during their missions. Similar applications might include smart bombs, cruise missiles, drones, and submarines.

The second application envisioned by DARPA was a Pilot's Associate. Imagine yourself at the controls of an advanced $20 million fighter like the F15 streaking along at 1,500 miles per hour. There is very little time to recognize or respond to a hostile event. This problem is typical of modern high technology weapons where there is a poor match between the capability of the weapon and the ability of the human to respond. DARPA's experimental solution is a Pilot's Associate, a collection of expert systems designed to help humans deal with this type of problem. It is capable of interfacing with humans in a natural way such as human speech and graphics. With reasoning, the system helps the pilot monitor the changes in the most pressing threats as his mission evolves.

The third DARPA application was a Battlefield Management System (BMS). It was intended to assist officers who must make critical decisions under levels of uncertainty. The current command and control systems in use, such as the World Wide Military Command and Control System, are rudimentary and very unsophisticated. A new BMS would offer new capabilities and go into the field with our forces. For example, it might be integrated into a complex naval battle group defense system. It could display, in real time, a detailed picture of the battle area, including the enemy plan for battle, the disposition of allied forces, electronic warfare environments, strike plans, and weather forecasts. Such capabilities simply are not available today.

Research activities under the SCI program are taking place on a broad front in the U.S. The shifting emphases on applications can already be detected at DARPA. Whether the results from this defense research initiative will have the desired effect in a timely manner in commercial markets remains to be seen.

Strategic Defense Initiative

On March 23, 1983, President Reagan suggested that it was possible to render nuclear weapons "impotent and obsolete" by creating new defensive weapons. This concept led to the creation of the Strategic Defense Initiative (SDI), popularly known as the Star Wars defense. Under this initiative, the government could expend as much as $30 billion for research and development funds pursuing several key technical program elements, including systems such as battle management, sensors for surveillance, acquisition, tracking and kill assessment, directed energy (laser and particle beams), and kinetic energy (such as rocket interceptors and electromagnetic launchers).

Most of these elements are closely related to, and heavily dependent upon, information technology. Thus, there is considerable interplay between the Defense Department's SDI office and DARPA. In fact, some DARPA SCI objectives, such as research in gallium arsenide, are now funded under the SDI program. And, the SDI office looks to DARPA for a wide range of research in AI, parallel processors, and distributed computing. Further, both programs are closely coupled since they are both managed by officials who report to the office of the secretary of defense.

The SDI research plan has created a political disturbance in the relationships among the major international powers, as well as in domestic circles. Scientists debate which parts of the program are feasible in the short term; political scientists question whether the proposal is more or less likely to provoke a nuclear war. The parliaments of the Western powers argue intensely whether their countries or NATO should join the United States in supporting SDI research projects.

The invitation to participate in SDI has been particularly awkward for Japan; legal and constitutional prohibitions constrain the Japanese in areas such as collective security arrangements, storage of nuclear weapons, and military deployments in space. However, the United States is anxious to exploit Japanese strengths in gallium arsenide, optoelectrical interfaces for signal processing, and some electronics packaging technologies. Responding to an invitation from the United States, Japan, in 1987, expressed willingness to work on some SDI projects. However, the Japanese government is proceeding cautiously toward implementing that agreement.

One drawback of SDI is that the ability to articulate or satisfy its software reliability requirements is extremely difficult. Some think it impossible, and a willingness to place confidence in automated defense systems with unproven reliability is considered a great, if not impossible, act of faith by many scientists. Many fear that such a system, which might incorporate an automatic launch on electronic warning mechanism, is too dangerous to build. One group argues it is unconstitutional, because it could preempt a presidential decision that is required by law. Others wonder how we will, in any rigorous manner, test or prove that the systems can meet their reliability demands under actual conditions of war. Consequently, the computer science field has seethed with controversy over this issue since the first SDI concept, Ballistic Missile Defense, was introduced in the early 1960s. Indeed, many software people, who have experienced failures in software systems that had previously worked for years, believe that an attempt to deploy SDI will result in an accidental exchange of nuclear weapons.

Further, satisfying the requirements in the original software specification may be beyond our current technical ability, and

this condition might exist for the foreseeable future. Nevertheless, there are other approaches (such as distributed battle sta-
tions) that we *might* be able to use to meet the requirements.
Also, intensive research into how to design and build systems
that cannot be tested could lead to progress or breakthroughs.
Alternatively, less than perfect systems can also affect strategic
and political decisions since they affect the uncertainties of our
enemies.[69] These political issues are closely related to the technical complexities.

Tensions between concerned scientists who have been vocal in
their criticisms and the government were heightened by remarks
by Donald Hicks, former under secretary of defense, who said,

> If they want to get out and use their role as professors
> to make statements [about SDI], that's fine, it's a free
> country. But freedom works both ways. They're free
> to keep their mouths shut ... [and] I'm also free not
> to give the money [for research support] ... I have a
> tough time with disloyalty.[70]

Disloyalty is a poor choice of words and certainly inappropriate,
and these remarks have been since disavowed by the Department
of Defense. Nevertheless, they do illustrate some of the bitterness
between the two communities.

Even with all of its problems and shortcomings, SDI is a huge
research program in all the technologies. We would expect both
useful research results and commercially significant products, not
only in software, but in architecture, optoelectronic technologies,
and energy sources. Such results should have a significant impact
on information technology.

Still, the efforts of the United States in the Technology War
may or may not benefit from this massive investment in SDI. On
the one hand, the technological achievements that we can expect
may be profound. But, on the other hand, the SDI investment
represents yet one more major investment in "big" science, such
as the great accelerators of physics. This investment is at the
expense of large numbers of smaller programs in science, usually

small and independent groups that work in a wide range of potentially useful problem areas and that often produce important breakthroughs.

Although we can count on SDI being a good process for developing existing ideas, will it serve to foster the next generation of creative ideas in the Technology War? Ordinarily, we would expect an affirmative answer. The field should benefit from such a focussed investment: with clearer objectives, there is less duplication of effort. Also, the massive programs tend to make it easier to assemble critical masses of people and the expensive capital equipment that is necessary to conduct productive research in information technology. But it is not clear whether the SDI research groups will have the charter or the freedom to spin off their technology into the commercial sector. In contrast, NASA had as part of its charter specific objectives for the commercialization of the technologies it develops under its focused space program.

Furthermore, pouring more money into the Defense Department serves to strengthen its hand at the expense of other agencies. At a time in our technological development when even greater coordination among agencies is needed, SDI further dilutes the role of the rest of government in the Technology War.

Special Department of Defense Programs

One of the largest problems in the Department of Defense's use of high technology is in the field of software. This is no surprise, since within the Department of Defense there is a 12 percent annual increase in demand for software, yet there is only a 4 percent annual growth in DoD's ability to supply software. Further, the software delivered by contractors to the Department of Defense often does not meet the department's specifications, and cost and time overruns are the rule rather than the exception.

Other software-related problems confronting the Department of Defense include technology transfer, the huge number of different computer languages and dialects, the reliability and adaptability of software, the dismaying slow rate of transfer of modern software technology into industrial and military practice, and the increasing technological complexity of computers and software that are required to meet future needs.

Some of the software cost overruns we mentioned may, in fact, be caused by technology transfer problems. It is difficult, for example, to move technology embodied in an algorithm into software. Algorithms, that is, procedures for calculating a desired quantity or result, can be very complex. Embodying them in computer programs (software) can be expensive and require many man-years of programmers. Exploiting progress in research laboratories in DoD computer applications is not an easy task; Defense lags the industrial sector in this process by up to 10 years. In many cases, some DoD sources estimate that technology transfer of software from the research laboratory to standard defense application takes fifteen years or longer.[71]

In 1975 a major software problem existed in that the Department of Defense could count some 400 different computer languages and dialects, including assembly languages, in use for mission-critical applications. This huge number had an awesome effect on the department's ability to maintain software, given its limited resources. The lack of standards, for example, compounded training, programmer productivity, and software maintenance problems. DoD decided to establish a common set of requirements for high-order languages and move toward a single high-level language. To get the broadest possible coverage and as up-to-date technology as possible, the Department of Defense sponsored an open international competition in language design. Four competitors offered alternative designs and, by 1979, a final choice was made. The new language was named "Ada" in honor of Augusta Ada Byron, Countess of Lovelace. (The countess, who worked in the nineteenth century with Charles Babbage, the father of the digital computer, is credited with being the world's first programmer.)

Ada is a polished language which supports many features that are important for complex military applications. Some of these features include real-time processing, modularization techniques, special treatment of unusual conditions, concurrency, and hardware interfaces. Ada was designed, in fact, to be easy to use in embedded systems, that is, weapon systems that depend upon mission-critical computer systems.

The Department of Defense controls both the Ada language and its implementations, through a Language Control Board and a validation procedure for its compilers. There are over 50 validated Ada compilers that run on many common computers. However, it requires more than a compiler and a language to help DoD improve the productivity of computer programmers. A set of common interfaces is also required so that tools written in an Ada environment can communicate with the underlying operating systems. Such an integrated sophisticated environment is now being built by DoD to support the use of Ada: the Ada Program Support Environment.[72]

Defense has done much to support and promulgate the use of Ada. Through its involvement in other programs, pressure is now being exerted on the military organizations that develop operational software to use Ada in current programs. One such effort to promulgate its use is the STARS program.

STARS is the acronym for the Software Technology for Adaptable Reliable Systems program. The program is intended to improve software productivity while offering greater reliability and adaptability of software. The actual activities to be included under STARS have varied; the program was started in 1983 but procurements were not funded until 1985. The early objectives included developing metrics for the measurement of software productivity, improving DoD's business practices related to software applications, and establishing software engineering environments and methodology. The funding through 1990 was set at $300 million, an amount equal to that of the first three years of SCI.

The Software Engineering Institute is a research facility supported by the Department of Defense. It is located at Carnegie-Mellon University. The role of the institute is to accelerate the transfer of modern software technology into industry and defense practice. About 60 percent of the institute's budget goes into technology transfer, 10 percent into research, 10 percent into education, and 20 percent into engineering. Their current work includes the study of the nature of technology transfer, development of a master of science curriculum in software engineering, the analysis of the reusability of software, and the development of automation and system design concepts. They are also assisting the Department of Defense in its development of Rights in Data clauses for use in its major procurements that involve large quantities of software.

The goal of introducing software engineering courses into the computer science curricula may turn out to be difficult for the Software Engineering Institute for both political and technical reasons. Some universities, for example, have been reluctant to teach software engineering techniques since much of the material is managerial and nontechnical and many of the methods are not technically sophisticated, it has failed to meet the schools' engineering standards. (In one case, a school offering a degree in software engineering was pressured into renaming the program by the state Society of Professional Engineers because the academic department did not have any registered engineers on their faculty.)

Another program supported by the Department of Defense is the Very High Speed Integrated Circuits program (VHSIC). It was organized in 1980. VHSIC is an integrated circuit effort driven by military systems needs for the late 1980s. The overall objective of this large program is the development of a range of IC technology for introduction into future military systems. To realize this goal, the planned VHSIC deliverables include a family of chips to perform the most commonly occurring, computationally demanding functions across weapons systems in all the military services; CAD techniques for the design of integrated circuits; pilot production lines to manufacture the family of chips, as well as custom chips; and systems prototypes to demonstrate VHSIC capabilities. The earlier successes of VHSIC are credited with the spawning of the SCI and STARS programs.

National Science Foundation

The National Science Foundation is an important government organization whose mission is to support basic scientific research. In thirty years NSF has developed into an important research funding source in science and engineering, although engineering applications are definitely of lesser importance than basic research at NSF.

The official goals of NSF in fiscal year 1986 were to increase the support for research in problems related to long-term American economic competitiveness (such as manufacturing automation), increase the support for programs that are key to the basic infrastructure of academic research, support quality programs in science and engineering education, support the Presidential Young Investigator Program, and reorganize the Engineering Activities of NSF. In fiscal year 1987, the NSF budget request was $1.6 billion. About $105 million went for computing. This included a 10 percent increase for mathematics and physical science and a 14 percent increase for engineering. In fiscal year 1987, one of the special programs for NSF is computational sciences and engineering. About $50 million was earmarked for advanced scientific computing facilities under this program.

Historically, computing has not been effectively organized at NSF; it was widely scattered and received little attention from management. But NSF has changed with the formation of a new directorate, Computer and Information Science and Engineering (CISE) in 1986. This is one of the five research branches of the foundation. Thus, the field of information technology is now suitably represented in the management structure of the foundation. The main research directions of CISE include the study of parallelism, automation, robotics, intelligence systems, microelectronics, supercomputers and networks. The research in the directorate is managed by four divisions: Computer and Communications Research; Microelectronic Information Processing Systems; Advanced Scientific Computing; and Information, Robotics, and Intelligent Systems.

We are delighted to see such innovation at NSF after years of neglect. Its new director, Erich Bloch, has a farsighted view of what is needed to improve the state of U.S. engineering, not only within the foundation, but at universities and other research establishments. There is a real need for these activities to succeed in America.

Some observers have tended to dismiss NSF and CISE as small and unimportant sources of funds when compared with DARPA and the rest of the Department of Defense. Although such reasoning is numerically correct, it ignores some important points. For example, NSF has funded seminal work in computer science over the years, and by the judicious use of funds, it *has* achieved great leverage. For example, one of its greatest successes was the Coordinated Experimental Research Program. The funds from this program provided the necessary encouragement to university faculty to cooperate with each other in the acquisition of equipment for the experimental side of academic computer science research.

In addition, NSF is still a major funder of basic research in the United States. When all of the computer-related activities are combined, NSF is spending a significant number of dollars, almost comparable with the levels of support provided by DARPA. The Reagan administration obviously is happy with the revitalization at NSF; the White House has expressed its intentions to support for the foundation's plan to double its budget for computer science research over the next five years.

Department of Energy Programs

The Department of Energy (DoE) supports a number of programs in energy-related work. Most of these are performed at national laboratories such as the Argonne National Laboratory, the Brookhaven National Laboratory, the Lawrence Livermore National Laboratory, the Los Alamos National Laboratories, the Oak Ridge National Laboratory, and the Sandia National Laboratory. The major categories of research in DoE's program are the basic energy sciences, high energy and nuclear physics, health and environmental research, and fusion energy.

Much of DoE's work involves computers and computation, and its category of basic energy science includes applied mathematics, statistics, and computer science, as well as techniques for numerical methods and high-speed computation. One particularly relevant activity, the DoE Applied Mathematical Sciences Research subprogram, consists of advanced computation and supercomputing research. The advanced computation program is dedicated to providing computing through computer networks, to supercomputers. This service is made available to DoE contractors in universities, industry, and DoE national laboratories.

In order to encourage the use of supercomputers, DoE provides open access to such machines in universities. However, such access is frequently complicated by the presence of scholars from communist countries. For example, most major campuses have visiting scholars from the U.S.S.R, China, and Eastern bloc countries. These visits are often facilitated by the Department of State. Thus, when DoE and, presumably, parts of the Department of Defense are seeking to provide access to university supercomputers to approved people, they find that they must prevent access to potential adversaries. Controlling access is especially difficult to enforce because most of the research is not classified in the traditional sense and conventional restrictions cannot be employed. Nevertheless, there is a presidential directive that states that restrictions (enforced through visas or otherwise) *may* be placed on unclassified federally funded work. Such control is unavailable overseas and the government's problem is compounded in Europe. For example, the Department of Commerce recently asked the University of London Computer Centre to guarantee that a Cray IS/2200 supercomputer would not be exploited by the thirteen Eastern bloc nations. The university, of course, balked at such restrictions.[8] Such resistance undoubtedly will provoke a reaction on the part of DoE and Defense and interfere with the future sale of American supercomputers to uncontrolled environments in Europe.

Among the laboratories that receive funding from the Department of Energy, the Lawrence Livermore Laboratory deserves special note. It has been the principal supercomputer user in the world and has installed the first version of generations of supercomputers. Indeed, Livermore is considered a prestige user and it is a matter of great honor for a supplier to install a prototype computer at their facilities. As a result, the serial number 1 of almost every supercomputer ever manufactured has been delivered to Livermore.

Perhaps few other agencies are as successful as DoE in technology transfer. Since the atom bomb projects of World War II, it and its predecessors have successfully developed high energy weapons starting with just the basic physics and carried its work to its completion in the advanced engineering of deliverable weapons. Its skills in technology transfer were honed in real warfare, and it continues to work under a mission requirement that demands high standards of reliability and accuracy. Its mission demands that it manage its technology transfer problems with success, and it does.

Support Of Computing Research

In summary, it is useful to examine the entire American funding pattern in the computing field. Table 7.1 illustrates the funding pattern by research area for fiscal years 1983 to 1985, and Table 7.2 breaks it down by research area and the particular agency.

Clearly, federal spending for research and development in the areas shown in the tables is rising, with most of the funding coming from the Department of Defense. Also, the percentage of support supplied by DoD has been increasing, reflecting the growing emphasis on defense engineering research and development by the Reagan administration. For example, research funding from the Department of Defense has recently risen from two-thirds to three-fourths of the total government research budget. And even though the Department of Defense's support of basic research fell far behind its increases in the more applied defense technologies,

Table 7.1 Research Area Funding in Millions of Dollars[73,74]

Research Area	FY83	FY84	FY85
Computation	15.7	16.8	18.0
Architecture	19.0	28.9	50.1
AI and robotics	30.3	43.8	63.1
Software	26.2	40.3	49.6
VLSI design	24.2	27.0	28.5
Data management	10.0	8.8	9.9
Theory	6.2	6.9	9.1
Networking	34.1	45.4	54.5
Performance	6.6	6.2	7.0
Total program	$172.3	$224.1	$289.9

Table 7.2 1985 Funding by Research Area and Agency in Millions of Dollars[73,74]

	DoD	NSF	NASA	DoC	DoE	Total
Computation	5.3	1.1	1.4	1.2	9.0	18.0
Architecture	32.9	4.6	4.6	0	8.0	50.1
AI and robotics	46.9	6.6	7.0	1.6	1.0	63.1
Software	39.4	7.1	0.9	0	2.2	49.6
VLSI design	23.9	2.6	2.0	0	0	28.5
Data management	4.5	2.8	1.7	0	0.9	9.9
Theory	4.4	4.4	0	0	0.3	9.1
Networking	29.0	17.3	2.2	5.0	1.0	54.5
Performance	0.3	4.6	1.1	1.0	0	7.0
Total program	$186.6	$51.1	$20.9	$8.8	$22.4	$289.8

such as those associated with SDI, Defense is the largest source of new basic research in the government.

What is the proper amount for a country to spend on research and development? The natural way to calculate that amount is as a percentage of the GNP or of the annual budget; a number close to 3 percent of the GNP seems about right. During a period of its dominance in the international commercial markets, the United States spent only slightly under 3 percent in the mid-1960s, but the figure declined during the 1970s. In the first half of the 1980s, the figure rose from about 2.3 percent to 2.7 percent. The Japanese figure has risen from 1.5 percent in 1962 to about the same percentage as the United States in 1984, that is, about 2.5 percent.

Two factors must be emphasized. First, U.S. spending is dominated by defense requirements. The percentage of military research and development to total research rose to 72 percent in 1985, as compared to about 47 percent in 1978. (Of course, Japan does not spend any significant percentage of its GNP on defense.) The result is that our principal high-technology competitor, Japan, spends most of its research and development on commercially relevant technology. Thus, after defense factors are discounted, both countries have comparable efforts in the commercial sector.

In that commercial sector there are some exciting companies that conduct high-risk research. These include the giants, such as IBM and AT&T, and some of the new consortia that have been formed to offset the high cost of capital in this research-intensive field.

8

United States Industry

We have chosen to confine ourselves in this chapter to the U.S. *information technology* companies. Nevertheless, most of the aerospace companies and a host of other companies in America, such as banks, are actively engaged in computer science, artificial intelligence, and other advanced computing research projects; the roster of companies performing important and innovative research include TRW, McDonnell Douglas, Citicorp, and General Motors, to name but a few. Still, it is the information technology companies that have the greatest influence on the results of the Technology War. Of these companies, the one that dominates the industry is IBM.

International Business Machines Co.

Some people find it difficult to separate computing and IBM. To some, IBM products exemplify the ultimate in technology — the science fiction beast of 2001. Few remember that the popular word for computers in the late 1950s was UNIVAC (for Universal Accumulator). Still fewer realize that IBM's equipment is considered somewhat inadequate in the computer science field, that is, the advanced environments found in research laboratories.

Nevertheless, IBM is *the* major force to be reckoned with in the entire field of information technology; is an occasional innovator in technology, manufacturer, and process techniques; is the first to establish and market an upward compatible and integrated product line; and, certainly, is one of the most effective users of legal talent in this century. Its willingness to take commercial risk in the marketplace, such as with the IBM 360 Series, is probably its most spectacular contribution to the field of information technology.

Of course, IBM has made substantial technical contributions, too. From the MARK I computer constructed at Harvard, to the numerous computational contributions in World War II, to new packaging technology, IBM's presence has been felt. Indeed, it is IBM's introduction of a series of manufacturing and packaging technologies that has kept the IBM-plug-compatible manufacturers scrambling to keep up (and mostly lagging). By staying aggressive, and exploiting its splendid marketing machine, IBM has become the de facto industry standard in both hardware and software.

The company's reputation and its record as a quality supplier are well established. This reputation contributes to its astonishing position in the marketplace, a position that has turned IBM into a national asset with important relationships to national security, industrial and regulatory policy, international economics, and the balance of trade.

IBM's Marketshare

IBM *is* large. It is not simple to comprehend a company with over a quarter of a million employees, whose annual sales were $50 billion in 1985.[75] Comprehension is even more difficult considering the erratic nature of the trade and public reports on the company's marketshare. For example, the Gartner Group has reported that in 1985, IBM held an almost 80 percent marketplace share in mainframe computers.[76]

In the important vertical markets, such as banking, discrete manufacturing, transportation and utilities, and government, the Gartner Group reports that IBM's marketshares were 95, 87, 87, and 85 percent (based upon numbers of installed units), respectively. *Datamation* tells us that IBM's 1984 software revenues of $3.2 billion were greater than the sum of all other software suppliers' revenues and roughly 20 times greater than the largest independent software vendor.[77]

Marketshare, which is one way we measure the strength of a company and its potential influence on industrial and national policy, has a constantly shifting definition — especially as a field expands. Marketshare is important; it sets buyer patterns, establishes standards (or nonstandards), and may even create new markets, such as personal computers. And, in this case, IBM dominates the markets, both domestically and in the international sector. Malik wrote, for example, "IBM is not just a major international company in the area of computing, it is the international environment."[78] In 1984 IBM Europe's sales represented almost 40 percent of the total data processing sales in Europe.

Since about 1971, IBM's World Trade Division has generally outperformed the domestic sector in profits, and in 1985 World Trade had income comparable to the IBM domestic sector income. This fact certainly has had a favorable effect on the U.S. balance of trade figures.[75] It is unlikely that the U.S. Department of Commerce would support any government attempt to weaken its competitive posture, its technological lead, or its ability to dominate the use of the technology. Conversely, the Department of Defense certainly would act to deny IBM technology to our adversaries.

Given the size of IBM, it is not surprising that it has drawn the fire of the Justice Department's antitrust guns. With all these data demonstrating that IBM is a dominant force in the world industry, let alone the United States, however, it is easy to see why the United States government has abandoned its antitrust action against IBM. Although there is no other tool besides the antitrust laws that could possibly moderate IBM's behavior *if* it was or is abusing its market position, the Reagan administration made a conscious decision to abandon claims against IBM in the interests of national security, technological advantage, and international competition. Such a decision may not offer cheer to IBM's domestic competitors, but it was the right strategic move in the Technology War.

In a rapidly expanding and innovative high-technology field, the only way to keep or capture market share is through research and development and the propagation of the research results into products. As the largest company in the field, IBM research deserves some attention.

IBM Research

IBM has several important research centers, located at Yorktown Heights, New York; San Jose, California; and Zurich, Switzerland, with a total of about 4,000 researchers. The budget is hard to estimate, but according to IBM it spent $4.7 billion on research, development, and engineering in 1985.[75] IBM executives have said that the company spends 7–10 percent of its gross revenues, the equivalent of $1.5 to 2.0 billion, on just research and development. Of this, probably 10 percent is spent on pure research. About $3 billion is spent on engineering.

IBM research has three official goals: to determine the fundamental limits of present technology, to develop new technologies, and to conduct the highest quality scientific research.

What hath IBM research wrought? In hardware, it has given us quite a bit. One of the greatest contributions of IBM research was the 350 RAMAC, the first commercial magnetic disk. This invention had such a profound effect that in 1984 it was declared an International Historical Landmark by the American Society of Mechanical Engineers.

Since then, IBM has remained at the forefront in storage technology; for example, RAMAC could store 2,000 bits of information per square inch, whereas today's IBM 3380 disk unit can store over 22 million bits in the same area. Silicon Valley may be a place where computers and microprocessors are considered to be the major business, but the dollar volume of storage components similar to the 3380 that are produced in that area matches those of the chip makers.

A substantial individual contribution in software on the part of an IBM employee was the work of Ted Codd, who invented the relational database concept — a novel approach to organizing databases. Based on Codd's concepts, the San Jose Laboratory developed System R, the first relational database system, and within ten years after that, there was a vigorous and growing relational data base industry. Codd's work was recognized by his profession; for this accomplishment he received the ACM Turing Award, the highest technical honor in the computing field.

IBM's work in surface physics and chemistry also has been at the leading edge of technology In 1986, IBM was incorporating 1 megabit chips in its high-end products, keeping pace with its Japanese competitors. Yorktown Heights was experimenting with 4 megabit dynamic RAM chips. IBM is also at the very leading edge in microelectronic research with its use of the scanning tunneling microscope. With it, we can "see" individual atoms. Recently, at Yorktown Heights, an experiment was conducted in which there was direct observation of electrons traveling through a semiconductor without undergoing any collisions. This is known as ballistic transport and is considered to offer the highest possible speed for an electron in a semiconductor. With such technology, it may be possible to build much faster electron devices than previously anticipated. Similarly, IBM is one of the companies at the forefront of superconductor technology, another field that offers extraordinary opportunities for computing.

One of the personnel innovations at IBM research was the creation of the IBM Fellowship, awarded in recognition of an outstanding contribution to the company. A Fellow is relieved of all normal corporate restraints for five years. The program was begun in 1963, and there have been ninety-two Fellows (only twenty-eight are in research), or 0.03% of the employees, making them an especially elite group. It is an IBM Fellow, K. Alex Müller, who is one of the pioneers in superconductor research.[79]

Curiously, IBM research results, particularly in computer science, seem rather unexciting. Universities, Bell Laboratories, and Xerox PARC appear more productive in research and with smaller budgets. Why? Perhaps, there is more exciting work, not being disseminated, which has not appeared in the product line. Is it the lack of graduate students? Bell Laboratories does not have them either. Is it size? It is very difficult for good ideas from the research division to be applied, in a timely fashion, by the operating divisions to products of IBM. For example, IBM Fellow John Cocke invented the idea of reduced instruction set computers (RISCs). Yet it took five years to produce the first IBM RISC machine, the IBM RT/PC. Does bigness preclude innovation? This seems to be the case regarding AI

> IBM, on the contrary, had a long ... checkered history of official skepticism about the whole subject of AI. Yorktown Heights, the largest of IBM's research centers, could fairly be described as censorious of, if not downright hostile to the idea.[20]

Of course, the problems in innovation and technology transfer are common to large companies in the United States. These companies have a difficult time supporting entrepreneurial activities because they cannot offer the necessary rewards to their employees to justify the risks normally associated with start-up activities. We have seen this same phenomenon in companies such as Xerox and Bell Laboratories. Innovation is ordinarily the result of risk — risk that large companies will not or cannot afford to take.

AT&T

Alexander Graham Bell invented the telephone in 1875. AT&T, his company, then grew for some sixty years until it constituted a regulated monopoly of the U.S. telephone system.[80] In fact, the United States had one of the highest quality and most reliable telephone systems in the world. Nevertheless, Bell's size and the rash of competitors, spurred by new technology, led to a growing number of legal clashes. Finally, in 1974 the Justice Department felt compelled to act. It filed an antitrust suit against AT&T. After eight years, the government accepted a settlement that allowed AT&T to retain long-distance service and manufacturing but required it to divest the 22 regional Bell operating companies. In exchange, AT&T was granted the right to enter the computer field. This was a momentous decision in both the communications and computer industries; both industries had displayed signs of a growing dependence upon each other and AT&T was clearly a company with a significant potential to exploit that growing relationship.

Years before the government's antitrust suit against AT&T, the FCC had conducted several inquiries of the computer and communications business to determine whether the rapid development of computer technology and computer networking might be constrained or whether innovation was being stifled by the manner in which the communications industry was being regulated. After carefully studying the issue, the FCC decided that communications companies would be precluded from participating in the data processing services business. This unusual decision was reached despite their conclusion that

> there is similar [virtually unanimous] agreement that there is a close and intimate relationship between data processing and communications services and this interdependence will continue to increase. In fact, it is clear that data processing cannot survive, much less develop further, except through reliance upon ... communication facilities and services.[81]

What is surprising is only that it took ten more years for the technical realities to ultimately overcome the political necessities and regulatory inertia of the FCC.

Divestiture has had its advantages and disadvantages to the consumer. The major advantage is the increase of competition, which fosters the faster introduction of new technologies and services; another is the substantially lower long-distance rates. The major drawback is that there is no longer one entity with end-to-end communications responsibility for telephone service. Further, deregulation has led, at least for the moment, to more expensive residential service. It also has led to a possibly permanent decline in the quality and reliability of service. That degradation in reliability may have adverse consequences on our national security.

In their new business, AT&T's initial computer products have not been well received; perhaps we must give the company time to adjust to this competitive business. Other major corporations such as RCA, Philco, Xerox, and General Electric have also found computing an impossible field to stay in, let alone to enter.

Nonetheless, there are good reasons why it is important for AT&T to succeed in both computing and communications under divestiture: It is only one of a handful of companies in the world that is in the grand marshall class in the Technology War. It has extremely important connections to national security, both in an operational and a research sense. And it serves as one of the few research and product alternatives to NEC and NTT in Japan. Perhaps no more telling indicator of its importance exists than Japan's recognition of the importance of AT&T and Bell Laboratories. Japan has modeled its national communications company and its Electro-Technical Laboratories after AT&T and Bell Laboratories.

Bell Laboratories

Bell Laboratories has created a standard by which all other such laboratories are measured. It too represents a great national asset that has served America well in the early skirmishes of the Technology War. But it hasn't been cheap! AT&T spent over $1.5 billion in 1981 to operate the laboratories.

The year 1981 was the last one in which the traditional laboratories could be evaluated without taking into consideration the effects of the divestiture. That year, Bell Laboratories had a budget of \$1.63 billion and some 24,000 employees. Its accomplishments included over 300 patents issued to laboratory employees, almost 6,000 presentations and papers generated and 89 awards won.[80,82]

From the beginning, Bell Laboratories made substantial contributions to technology. For example, the invention of the transistor took place at Bell Telephone Laboratories in 1947. It is a testimony to its importance that everyone has heard of the transistor and how it replaced the earlier vacuum tube technology. But, typical of U.S. technology transfer problems, the initial reaction on the part of American industry to the discovery of the transistor was indifference. It took another six years before transistors appeared in commercial products and many of these were in Japanese radios.

By its very nature, Bell has been an important participant in the computer field. In the 1970s, very important operating systems work — the development of the UNIX operating system — was done at the laboratories by Ritchie and Thompson. Proving once again that technology transfer is difficult to plan, the UNIX project was not established to meet any business objectives, nor was it carefully organized by management. It was a serendipitous consequence of having bright people and useful equipment at the right place and the right time.

For example, Thompson, experimenting with an unused DEC PDP-7 computer, developed a personal computing environment. This caught the interest of his colleague Ritchie, and the two men began to collaborate. From such a humble beginning, the UNIX operating system was born. At present, the system is used extensively by thousands of people. For their work, Ritchie and Thompson received the ACM Turing Award.

In his acceptance speech, Ritchie said that "Bell Laboratories has provided ... a rare and uniquely stimulating research environment."[83] He speculated that outside the environment of the laboratories it would have been impossible to develop UNIX, one of the few alternatives to IBM operating systems on mainframe computers.

The years 1982 and 1983, when divestiture was underway, were a transition period at the laboratories. Bell lost 4,000 of its staff to AT&T Information Systems and an additional 4,000 to Bell Communication Research. These are new organizations. AT&T Information Systems is a computer manufacturer, and Bell Communication Research (Bellcore) is a central research laboratory for the seven regional operating companies in the United States. In fact, Bellcore, employing about 7,500 employees, is now the largest collective research activity in the United States. Its budget is about $1 billion and is devoted to computer science, solid-state physics, light-wave communications, and network research. Software development appears in all these categories and represents the largest activity in Bellcore.

The laboratories are now "running like hell to change the technology mix."[84] Bell Laboratories is seeking "killer technologies," those that kill off the previous generation of technology. An example is the pocket calculator and the slide rule.

The budget is no longer lavish; 1986 brought a salary freeze. Managers must now be alert for commercial opportunities. Seeking to encourage technology transfer, bonuses are offered to encourage the transfer of research ideas to the development groups. On the other hand, the openness of the laboratories has been reduced. More care is taken in the clearing of material for publication, slowing outside technology transfer.

There is an interesting case, illustrating the change at Bell Laboratories, of a young researcher named Narendra Karmarkar, who invented a significantly different algorithm for solving linear programming problems. Linear programming is an important combinatorial technique with applications to many economically significant optimization problems, such as those in resource management, transportation, and communication. A great deal of computational effort is spent on solving these linear programming problems. No cost-effective algorithm for many of them was known until a solution was developed several years ago by Khachiyan. Unfortunately, this algorithm was only of theoretical interest because, among other reasons, it was numerically unstable. On the other hand, the Karmarkar algorithm was well behaved numerically and it provided new and cost-effective solutions. Important linear programming problems were solved at record breaking speeds with this technique. However, under the new ground rules at Bell Laboratories, the programs and the implementation tricks used to achieve the results were not made available to the scientific community. Bell is using Karmarkar's algorithm in designing its Pacific Basin communication networks and continues to protect its proprietary position.

Clearly, Bell has decided that it risks losing a competitive advantage by revealing too much. Under the rules of the old Bell Laboratories, those that existed before divestiture, it is safe to say the findings would have been published and that technology transfer to other organizations would have been speeded significantly.

Digital Equipment Corporation

Digital Equipment Corporation (DEC) is the third largest computer company in the United States and currently is one of the most profitable in the field. It is closely managed by Kenneth H. Olsen, a visionary who identified the existence of the minicomputer market in the 1960s. In that period, DEC introduced the first minicomputer, which cost $120,000 in contrast to the millions of dollars charged for mainframes. By 1966, it was offering the PDP-8, the first mass-produced minicomputer, for only $20,000. DEC's sales reached $1 billion in 1977 and during the 1970s, its revenue growth was astonishing. Indeed, DEC revenues were growing at an annual rate of 86 percent. By 1982, DEC had reached $4 billion in sales, primarily on the strength of the VAX computer family. Its systems were used in most computer science research facilities, and its hardware was preferred over most in sophisticated digital communications systems. In 1985, DEC actually experienced greater growth in revenue (percentage-wise) than IBM, and its sales in 1986 exceeded $7 billion.

Nevertheless, in the mid-1980s, DEC encountered difficult

times. New computers were years behind DEC's development schedule. Repeating the same type of error made earlier by IBM, DEC ignored the burgeoning personal computer market: DEC had engineering talent, a highly automated and high-volume manufacturing facility, and significant sales and service facilities. It would have been well positioned to move into the personal computer marketplace; sadly, it arrived late in this market.

Still, DEC is well positioned in a most active segment of the minicomputer market. Sales and profits are exploding and DEC's short term prospects are excellent. In the long run, however it is unclear whether DEC will continue to thrive or be crushed from above by IBM's entry into their markets with the introduction of IBM products such as the 9370. Similarly, it may be outmaneuvered from below by more nimble upstarts such as Sun. Further, it faces an unclear threat from Unisys, which must consider DEC's markets an easier target than IBM's.

DEC equipment has an important role to play in the Technology War. As the supplier of equipment to most research laboratories, its products will have considerable influence on the research trends and results. Still, DEC has been distancing itself from its traditional computer science markets in order to gain commercial marketshare. We take its introduction of workstation products as a welcome attempt to reestablish its position in the research community.

Nevertheless, DEC has not had a consistent and clear notion of its own role in research. It has not produced notable research results and DEC's contribution to the Technology War effort is through its customers rather than through its own laboratories. It is interesting that DEC's successes, such as the VAX computer series, have been accomplished by concentrating on engineering development and not by focusing on research. DEC management must realize this because it has recently opened two separately administered research laboratories in Palo Alto (across the street from each other). Both are at considerable distance from Massachusetts, where the corporate decisions are traditionally made. DEC has ignored the lessons of history offered by Xerox, which suffers the same geographical disadvantage and has failed to overcome its own technology transfer barriers.

DEC serves of an example of a viable and independent alternative to IBM across most of the product line. IBM's other competitors in the U.S. are often considered together as a group, known collectively as the BUNCH.

The BUNCH Companies

Before the merger of Burroughs and Sperry Univac into Unisys in 1986, the computer companies Burroughs, Sperry Univac, National Cash Register, Control Data, and Honeywell were known as the BUNCH; an acronym derived from their names. These companies and DEC, as they existed in the past and as they exist today, have represented most of the serious non-IBM-compatible commercial automatic data processing competition in the United States. Together, the BUNCH shares about 18 percent of the mainframe hardware marketplace.

The BUNCH has evolved from a larger group of companies. In the 1960s, General Electric and RCA, along with the BUNCH, were known as the Seven Dwarfs and were considered serious contenders to IBM. As one source wrote in 1965, "The smart money for the short run is on General Electric and RCA."[9] However, General Electric and RCA left the scene early, leaving the only meaningful domestic competition to the other dwarfs.

Because the BUNCH represents a serious competitive alternative to IBM, what happens to the companies is quite important in the Technology War. Through competition, they provide some of the few incentives to IBM to innovate. Further, they are large companies with a substantial international presence. Each company is larger than the data processing division of any of its foreign competitors. With enlightened leadership and help from the government, they can still make important contributions in the Technology War.

Compatibility

Compatibility is an extremely important issue in data processing. The lack of compatibility means that users cannot take advantage of new, more cost-effective equipment or move to alternate hardware from other suppliers that will process their software. IBM compatibility is even more important since the market of IBM and IBM-compatible equipment represents the largest market. Moreover, this equipment is the favored choice for vendors of third-party software products.

Compatibility has been the bane of data processing since the days of Univac's 90 column punched cards and IBM's 80 column punched cards. In 1981, William S. Anderson, then chairman of the board of NCR, described systems incompatibility as the "curse of today's information processing world." Referring to a new system for interconnection of disparate systems, he said, "it can help the industry cut its way out of today's jungle of systems incompatibility."[85] Still, despite the "curse," the BUNCH, as part of their market strategy, wagered that IBM compatibility would not be critical in their specific niches. Those strategies have proved less than successful; none of the BUNCH holds a significant hardware marketshare.

The BUNCH companies find themselves under pressure on many fronts: IBM is pushing relentlessly across the entire data processing product line; the IBM-compatible suppliers, such as National Advanced Systems, are sweeping up the crumbs left by IBM from customers who prefer an alternative to IBM but who wish to stay compatible at the expense of the BUNCH; technology in younger and more nimble companies is being used to wrest marketshare in specialized niches away from the BUNCH; and the microprocessor and minicomputer manufacturers, such as DEC and Hewlett-Packard, are pushing up into their market niches from below. In the international sector, the BUNCH faces fierce price competition from the domestic suppliers. As a result, their combined worldwide share of revenues from mainframes declined from 26 to 18 percent between 1979 and 1983.

The BUNCH may have abandoned compatibility with IBM, but they are seeking to connect to IBM equipment through communications networking. In doing so, they may have adopted a strategy that recognizes their second-class position. As *Informa-tion*WEEK put it:

> These strategies seem to amount to an admission by the BUNCH that they've lost the battle for the heart of corporate data processing and are now switching to guerrilla tactics to chip away at the edges of sites where IBM computers control the main ... data.[86]

Evolution of the BUNCH

Before their merging into Unisys, Burroughs was known for its presence in financial data processing, while Sperry Univac concentrated in the traditional defense and communications markets. Both were seeking to develop vertically integrated products by establishing strategic partnerships and focusing on technical integration.

Actually, Burroughs had a reputation for superior hardware and operating systems. The company introduced time sharing, real-time operating systems, and several architectural features (stacks) in the mid-1960s. But, the company never mounted an IBM-like sales program. Depending on customer loyalty to preserve its marketshare, it slowly watched that marketshare erode.

Sperry Univac has a long history of supporting scientific and communications applications (the UNIVAC I was introduced in 1949). Univac also pioneered the use of transistors in 1956, and introduced innovative and important programming concepts in 1960. Nevertheless, about 60 percent of Sperry Univac's revenues were earned in the commercial electronic data processing marketplace.[87] Sperry Univac, alas, inherited the RCA mainframe base in 1971 and has suffered incompatibility problems ever since.

Together, Sperry Univac and Burroughs make strange partners with incompatible markets and product lines. They are going to have a hard time merging their businesses. It is, indeed, difficult to understand how Unisys, the new company replacing Sperry Univac and Burroughs, intends to rationalize the disparate product lines. The jury will be out for some time, but we don't expect much more success than Honeywell had when it absorbed General Electric's computer business.

NCR, founded in 1894, is the company that began it all by training T. J. Watson, Sr., the first chairman of IBM, in the art of sales. It entered the computer business in 1952 with the acquisition of the Computer Research Corporation. However, NCR did not become a serious competitor until 1967 when it introduced the Century series of machines. The company is rather strong in IBM-compatible communications equipment, custom microprocessors, and UNIX multiuser systems. Unfortunately, there does not seem to be an easy way to unify its product line of over 250 products in its specialty marketplaces.

Control Data Corporation (CDC) is in a lot of businesses, including computers, peripherals, finance, health, and day care centers. It has been an innovator in instructional technology (such as the Plato system), but such products have not experienced good market demand. It is considered a specialist in scientific computing products, and one of its greatest successes has been in the field of supercomputers. However, it is currently under substantially increasing competitive pressure from supercomputer manufacturers, such as Cray.

One of the few companies that obtained concessions from IBM in an antitrust action, Control Data Corporation has not, since then (1973), assumed the expected leadership in the field. However, CDC has made substantial contributions in the Technology War. It has been, for example, a major catalyst in the formation and support of important U.S. research coalitions, most notably MCC.

Honeywell acquired the General Electric computer base in 1970 and became, at that time, the second largest computer company in the world. Honeywell, specializing in manufacturing and control, became an even larger company when it acquired the Xerox computer base in 1976.[9] Its marketshare might have grown even more substantially; their H200 system did very well against the IBM 1400 in the early 1960s. In fact, Fishman has reported that IBM was so distressed by the unexpected market acceptance of the H200, that it actually advanced the announcement date of the System 360.[88]

Honeywell was no more capable of dealing with the competitive pressure from IBM in the 1960s than it is today. Indeed, Honeywell has formed a new partnership with NEC and Bull, a French computer company, and is no longer a mainframe manufacturer, relying on NEC for its source of hardware. Honeywell is one of the latest American casualties in the Technology War.

Hewlett-Packard

The story of Hewlett-Packard (HP) is one of a garage enterprise that succeeded beyond imagination. It serves as the ultimate example of entrepreneurial success and led, to a considerable extent, to the establishment of the Stanford Industrial Park and, ultimately, the Silicon Valley.

Because of HP's successful entrepreneurial root, it embraced a corporate policy intended to foster more such success and overcome the barriers raised by traditional technology transfer problems. Groups within HP are encouraged to design, develop, and then market their new product ideas. Thus, Hewlett-Packard provides its employees with the benefits of working for a large company, as well as the chance to learn entrepreneurial skills while they develop and commercialize their own ideas. This combination of advantages is usually found only in small startup companies. It is this entrepreneurial spirit that is of special interest in this book.

HP is both a scientific instruments and minicomputer company. It has been tremendously successful in scientific instruments, but it has faltered as a computer company. HP has seemed less sensitive to the differences between computing and other forms of electronics than its competitors. Higher technical management is drawn mostly from the older electrical engineering milieu. This group was raised in an era that lacked appreciation for the importance of software and the benefits of an integrated product line. As a result, HP has had varying degrees of success. When working on innovative devices, such as the first handheld calculators, the company has been successful. Certainly, in that instance, HP must be considered as the company that set the standards. Unfortunately for HP, they fell victim in these markets to the innovative pace of Japanese manufacturers. And, in computers, it has been an uphill battle all the time.

In spite of these difficulties, Hewlett-Packard is a company with strong financial resources, having $5 billion a year in minicomputer and software sales. It enjoys a fine reputation for service and quality manufacturing and has good technical people, particularly in the traditional electrical engineering areas. But the question remains: Can Hewlett-Packard be successful as a computer company? HP is proud of saying that it is now a computer company because the majority of its revenues come from computing operations, not from instrumentation. The important questions, though, are how well will its new Spectrum minicomputer series do in the presence of growing competition and what reception will its workstations receive in the commercial data processing marketplace?

Other Companies

We have examined many of the American corporate giants and the major players in computing, but there are other companies, including start-ups, that are quite exciting. Some of these companies are new and are at the very edge of technology; a few may become the industrial giants of tomorrow. Some may not survive at all or will disappear through acquisition. We discuss below some of the more popular and well-known companies in this class, such as Apple, Xerox, and Sun Microsystems. They all have an impressive history of innovation; their commercial accomplishments, however, have not been uniformly impressive in the marketplace.

Apple is one of the most recent American success stories. Perhaps no other company deserves the credit it does for establishing the personal computer as a real product (Apple II) in a viable marketplace. After only five years in this new marketplace, Apple's sales had reached $1 billion and its after-tax profit margins were about 13 percent. Apple's initial stock offering sold at twice what most technology companies expect in terms of price-to-earnings ratios, and a number of Apple employees became millionaires.

But Apple had problems in the early 1980s. The Apple III was a flop, perhaps because the market perceived it as a loser. McKenna suggested that the Apple III got off to a shaky start, and simply never met the expectations of the market.[89] Fortunately, the dependable Apple II, revised into the Apple IIe, was a great success. By generating 97 percent of Apple sales in 1983, the Apple IIe carried the company into the mid-1980s.

Steve Jobs had an opportunity to see for himself the improvements in personal computers possible with sophisticated software, graphics, and input-output devices such as the mouse. He was exposed to these technologies during a visit to the Xerox Palo Alto Research Center, where some of the fundamental research in these technologies was completed and demonstrated. Jobs first brought that technology to the marketplace in the form of the LISA. Unfortunately, the LISA was poorly timed for both of its potential markets, business and personal use: LISA was too expensive at the time to be labeled a personal computer, and since it was also not IBM-compatible, it was unsuitable for the corporate market.

The Apple Macintosh, however, has met with more success. It is aimed primarily at the personal market, and only secondarily at the business market. It is a low-priced machine with one of the best user-friendly interfaces available. Some of the most creative software in the industry has been written for the Macintosh, further enhancing its appeal. Apple also introduced a local area network and a superb laser printer to provide the ingredients for Macintosh office applications. This has broadly expanded the appeal of this machine in the business community.

Since the introduction of the Macintosh, Apple has remained a nimble and creative innovator in the field, one of the few that seems to have bridged the technology transfer gap. Having mastered the offense, Apple is now playing defense very aggressively by invoking the intellectual property rights laws in order to protect its proprietary investments. Its litigation is going a long way toward establishing the legal framework for the copyright protection of software. As a result, Apple is a critical firm to watch in the future in the Technology War.

Xerox entered the information industry with the introduction of the legendary 914 copier in 1959. Unlike its competitors, which required specially treated or heat-sensitive paper, the 914 used ordinary, untreated paper. It was probably the single most successful high technology product of all time, and great fortunes were amassed. In fact, the Haloid Xerox Company, which produced the 914, started as a $30 million company and by 1965, the successor company, Xerox, had revenues of almost $400 million and most of the marketplace.

Unfortunately, ten years later, the company's profit increases became single digit, and the Xerox marketshare in copiers fell from 80 to 40 percent. The reason was the competition offered by the Japanese, who captured over one-third of the American copier market. In fact, Japanese companies such as Canon and Ricoh and their distributors now have claimed the lead in the low-end copier market.

It is a familiar story — success led to market dominance and then to complacency. The small copier market was deemed too unprofitable for Xerox, and thus an umbrella effect was created for Japanese rivals. But Xerox ignored the threat from Japan and devoted its attention to potential U.S. competitors such as Kodak and IBM. It must have shocked the Japanese to discover their biggest competitor was ignoring them. It's no wonder that a Japanese competitor said, "Xerox is a piece of cake."[90]

Computers offered Xerox one way to diversify into other high technology markets. Seeking such opportunities, in the 1960s Xerox acquired Scientific Data Systems for $1 billion (in stock). Renamed Xerox Data Systems, the company was known for mini-computer systems and was an early supplier of timesharing products. But, after very poor experience in the marketplace, Xerox wrote off this investment in the computing industry for $84 million in 1975.

Since then, the story of Xerox's products and achievements in computing is one of limited accomplishment. On the one hand, Xerox Palo Alto Research Center has pioneered in important leading-edge research in workstations, networking, and user and programming environments. On the other hand, Xerox computer products continue to be poorly received in the market. The Xerox STAR, an early workstation, had slow user response times, an inadequate disk interface, and was overpriced. Perhaps Xerox's computer marketing practices and resources are weak. The STAR, for example, was poorly marketed and was not compatible with other mainframes.

Although Xerox has not succeeded in computing, it has been successful in the fight to staunch the loss and recover some of its marketshare in copiers. Xerox has accomplished this with outstanding new products, a revitalized marketing force, and a good working partnership with Fuji Xerox.

The tale of Xerox shows that success in one high technology field and marketplace, like copiers, does not necessarily carry over to another, like computers. Consider that despite IBM's success in computing, it has not been successful in the copier field. Nor has AT&T's dominance in communications been reflected in computing.

Sun Microsystems (Sun) is an exciting new company whose products center around personal workstations, primarily for the engineer or scientist, that feature a fast microprocessor, networking capabilities, and a high-resolution bit-mapped graphics display. Sun was founded in the early 1980s when, as a result of a recently passed tax reform bill, venture capital was widely available for high technology companies. Because their initial products were at the state of the art, Sun was able to attract more innovative designers and programmers than its competitors.

In fact, in 1982 Bill Joy, a Ph.D. candidate from the University of California at Berkeley, joined Sun. Joy, who had been the principal designer and implementor of the Berkeley 4.2 version of UNIX, a popular and widely disseminated operating system, provided the company with valuable and unique software expertise. This technology was incorporated into Sun's products, which were rapidly accepted by the academic research community, and the company grew dramatically.

By 1986, Sun had sold over 12,000 of these high performance systems and realized revenues of $310 million. Their typical systems cost from $8,000 to $70,000. Sun has been competing with Apollo for workstation marketshare, but it seems destined to surpass Apollo. Their next natural competitor is DEC. With this growth rate, Sun should be a *Fortune* 500 company in their fifth year.

One of Sun's weaknesses is its dependence on its suppliers. Sun's systems, for example, are based to a large extent on other suppliers' products: Motorola manufactures the microprocessors, others make the memory chips, and so forth. And, although Sun is a major force among UNIX users, there is competition, and many commercial customers want UNIX System V compatibility, an AT&T product. There are also threats from overseas. Many Japanese workstations have been announced, and some are expected to have prices lower than Sun's. Also, IBM has introduced the PC/RT workstation. Although it has poorer performance, IBM is expected to double and triple the speed of its RT. In response, Sun has adopted an innovation strategy to avoid giving either IBM or the Japanese a fixed target.

Sun is a favorite of ours because of its entrepreneurial background and aggressive approach to technology transfer, because this kind of company pushes the limits of technology, and because it forces others to do the same. It is the quintessential example of those high technology companies that are crucial in winning the Technology War.

Industrial Consortia

As we shall see in the next chapter, many of the Japanese companies in the information technology field are huge, vertically integrated companies. Many of the small, as well as large, U.S. firms are no match for a NEC or some of the *zaibatsu* with large electronics companies. So it is logical for U.S. companies to combine forces and share the risk in long-term research, even if such an idea is somewhat alien to our ideas about competition. Nevertheless, there comes a time to try radical ideas. With blessings from the Departments of Justice and Commerce, several strong and quite novel consortia have emerged. We describe below some of the more interesting experiments.

In 1982, William Norris, who was the CEO of Control Data Corporation, called a meeting of key executives of the major U.S. computer firms. He expressed serious concern that many American companies could not afford the resources needed to mount expensive high-risk research programs. He suggested that a combined research effort would spread the risk among a large number of companies and provide a large talent pool.

The timing was right. The international pressures were increasing, technology development was becoming more capital intensive, the companies were finding that they were losing their technical advantages, and the government was sending friendly signals. Also, Norris had an excellent candidate for the job, Admiral Bobby R. Inman, Jr. a former deputy director of Central Intelligence and past director of the National Security Agency.

Inman, a long time user of technology from his days in naval intelligence, had always enjoyed good credibility in the scientific communities. The scientists appreciated his candor, intelligence, and integrity. It was clear he was qualified to recruit and manage a group of computer scientists. He was also concerned personally about the United States' precarious domination of technology, and he was intimately familiar with the threat to the United States from the exploitation of high technology. With a military background, Inman suffered from no corporate biases, and he was well equipped to shepherd legislation such as the National Cooperative Research Act of 1984 through Congress. Inman's availability was probably a key catalyst in forming the new coalition proposed by Norris.

The Microelectronics and Computer Technology Corporation (MCC) was established by eleven American corporations and is located in Austin, Texas. MCC is a profit-making corporation specializing in advanced research and development. Funding is provided by the shareholders. Its companies pay a basic fee of about $1 million each and participate in at least one of the projects. The profits, if any, will come from royalties earned by licensing inventions. MCC spent approximately $65 million in 1986 against a research capitalization of about $700 million.

As of 1987, MCC's plans call for four general technology programs. (1) The AI work under the Advanced Computer Architecture Program is oriented toward extending the state of the art in the development and use of expert systems with new computer architectures and the use of new device technology to increase speed. (2) MCC's CAD program focuses on the development of system tools for the design of complex, high-density semiconductors. (3) The aim of the MCC packaging program is to develop cost-effective packaging technology capable of interconnecting devices in a minimum area with maximum electrical and thermal performance. (4) Finally, the MCC software program ranges from work in theoretical computer science to database management and documentation tools.

The future of MCC is unclear. One concern is that the company will not produce results in time to aid its members, who may then simply disappear or withdraw support. Secondly, MCC must conduct long-term research; results may not appear for seven to ten years, if ever. American businessmen are not noted for their patience and, further, some of the shareholders may not be in business in ten years. Also unclear is what success MCC will experience in technology transfer. MCC, after an effort of almost one year, put into place a master technology transfer licensing plan. But, it is too early to measure its effect. It is worth noting that Admiral Inman's biggest worry as MCC's CEO was not whether the technology would be developed. His greatest concern was: "Technology transfer — now the question is, will it get used?"[91]

Perhaps most unclear is MCC's future under a new leader. Admiral Inman has resigned to pursue other interests. Is MCC beyond the critical mass? Only time will tell.

The Semiconductor Research Corporation (SRC) was formed in 1982 in another response to the loss of microelectronics marketshare to the Japanese. At that time, the semiconductor manufacturers felt that one of the most pressing problems in the United States was the shortage of skilled electronics graduates with a research and development background. SRC was created to increase the amount of academic semiconductor research: it also provided considerable state of the art equipment to the universities.

SRC's approach of cost sharing for research and equipment made it possible for small companies without enough researchers to share the risk in highly advanced research. Between 1982 and 1984, SRC's budget approached $30 million; in 1986, its budget was $20 million, 90 percent of which was funneled to over forty universities. Research results accrue to all industrial participants and the universities have no publication restrictions. By 1987, SRC has already produced sixteen patent applications and has supported over sixty graduate students.

The Microelectronics Center of North Carolina (MCNC) is a partnership of educational institutions designed to perform basic research in microelectronics and to expedite its transfer into industry. MCNC has a $44 million industrial research facility in Research Triangle Park, North Carolina. It combines the resources of five universities and the Research Triangle Institute for the purpose of education, research, and development in the areas of semiconductor materials, devices and fabrication processes, computer science, CAD, modeling and simulation, and integrated circuit design.

Current projects are pushing the design and integrated circuit fabrication facilities to their limits. Among the key goals supported by MCNC are a vertically integrated microelectronics design system, a fast prototyping capability for building experimental architectures, and an advanced silicon wafer fabrication facility.

Perhaps the most interesting aspect of MCNC is its unique organization. It obtains tremendous leverage from the cooperating universities, the state of North Carolina, the federal government, and industrial participation. This type of synergism seems beneficial to all parties. There has been almost $100 million of new investment in North Carolina; over $40 million is from the state.

The consortia described above are representative of the significant, new, and novel experiments in American technology development. They represent the response of government and industry to the technology shocks of the 1980s, such as the Japanese capture of our high technology markets and the announcement of the Fifth Generation Project. The benefits that these consortia offer, such as cost sharing of high-risk research, are obvious. We endorse these experiments and hope that America will continue to support these cooperative endeavors.

9

Japan

Japan, among all of the technology warriors, is the one that excites the most concern and fear in the West. The Japanese are perceived as being competent, highly motivated, and a resourceful competitor. Indeed, from the devastation wrought by World War II, Japan has climbed out of its ashes, rebuilt itself, installed a new form of Japanese government (democracy), and adapted itself to the new world of competition governed by Western rules. Through hard work, farsighted policy, and a loyal workforce, Japan has succeeded in making itself a dominant economic power. Japan should be proud of these achievements.

One measure of these accomplishments is the fact that its exports are often among the best in the world, independent of price. When price is added as a consideration, Japanese products are unsurpassed. These exports have led to a rapid expansion of Japan's international marketshares, representing early victories in the Technology War. But, easy initial victories in any war, as the lessons of World War II taught, can be hard to sustain. It takes, at least, government support, either in the form of direct involvement, or, alternatively, a consistent, stable, and loosely regulated economic system. These attributes seem to abound in the Land of the Rising Sun where they are integrated together in a great national effort.

The Japanese National Effort

As discussed in Chapter 6, Japan has achieved its particular form of success by highly coordinated national efforts. The coordination was orchestrated by the Ministry of Trade and Industry (MITI), the master strategist that deserves the credit for Japan's modern industrialization. MITI carries the burden of sustaining, if possible, Japan's early successes in the Technology War.

Ministry of Trade and Industry

MITI, the grand marshall of the Japanese activity in the Technology War, has been attracted to information processing since the late 1960s. With amazing foresight, the ministry created the Machinery and Information Industries Bureau in which it concentrated responsibility for the national development of computers and automation. The bureau's work led to spectacular marketing successes in numerically controlled machine tools and robotics.

MITI encourages selected industries by supporting cooperative research and development on a wide industrial and academic basis. In the information technology field, the key companies that work with MITI in Japan are well known. These major corporations are Fujitsu, Hitachi, Mitsubishi, Matsushita, NEC, Toshiba, and Oki Electric; there are also smaller firms, such as Sharp.

The firms find it useful to participate in MITI projects for several reasons: MITI provides considerable funding for precompetitive research, it provides a check and balance on the firms' key competitors, the companies benefit from consortia-based research in capital-intensive programs, they have an opportunity to influence government policies and programs, and for new technologies and new markets, the partners can expect government protection for their fledgling activities.

Of course, MITI does make mistakes from time to time and not all firms slavishly follow its guidance. One of the most notable exceptions to MITI's policies was Sumitomo Steel's rejection of a MITI suggestion to accept a limited marketshare allocation in the steel markets. Sumitomo resisted MITI, until it achieved its desired compromise, in a case that received great notoriety in the press. Another example was NEC–Toshiba and Oki–Univac–Mitsubishi's refusal to go along with a MITI plan to restructure the computer industry.[92] It is interesting that NEC's rejection of MITI's plan has not interfered with the company's success. Indeed, NEC is one of the few computer companies in the world that has been successful despite its lack of compatibility with IBM.

Universities also participate in MITI programs (funding for the professors who serve on MITI advisory panels is provided by the Ministry of Education). The universities most often identified with MITI are the University of Tokyo, Keio University, Kyoto University, the University of Toyohashi, and the University of Tsukuba.

MITI Research Laboratories

MITI has alternate methods for satisfying the national research objectives of Japan. There is, for example, a large centralized and permanent research facility that is comparable in size and structure to a large national laboratory in the United States, such as the National Bureau of Standards. Another model occasionally employed by MITI, on a temporary basis, is to establish a national project, coordinated by committee, with the actual research conducted in either government or industrial laboratories. Yet another method is the establishment of a special purpose laboratory that supports a specific research objective. A complete understanding of MITI's role in research and development requires an examination of all of these models.

The Electro-Technical Laboratory (ETL) is a national facility located principally in Tsukuba City. ETL's work, funded and managed by MITI, is used to supplement the research and development results of Japanese companies or other government laboratories. ETL, when measured in American terms, can be viewed as a combination of the technical activities of the National Bureau of Standards with many of the scientific strengths of the Los Alamos National Laboratories.

The ETL research staff of 600 is organized into fourteen divisions, many related to information technology — fundamental science, materials, electronic devices, information sciences, computer science, computer systems, and automatic control.

ETL's results are disseminated to companies participating in MITI programs and usually become available nationwide. These results frequently appear in commercial products, and ETL is regarded as a major national asset in Japan. The MITI laboratories in Tsukuba City form the core of the scientific arsenal in Japan. It's no surprise that the 1985 Expo in Japan had science as its theme and was held in Tsukuba City.

There is also another research facility operated by MITI that deserves attention in any study of information technology. It is the Institute for New Generation Computer Technology (ICOT). The participating companies are obligated to give ICOT their cooperation and technical support, including key technical staff for assignments at ICOT. The Electro-Communications Laboratories (ECL) of Nippon Telephone and Telegraph and ETL, together with the companies, have furnished about 100 senior researchers who work at or manage the ICOT research center.

Research groups in the participating companies' laboratories have been assigned to track progress at ICOT and absorb the research results for their proprietary use. To facilitate this transfer of technology, ICOT researchers rotate back to their companies after three or four years. The researchers also routinely report back to their companies on progress of the project. Including the researchers at the outside labs, over 200 technical people are directly involved in the ICOT project.

Some comments on the rotation of researchers are in order. Ordinarily, such rotation is a valuable technology transfer mechanism; indeed, personal interaction is the best method for transferring technology. However, there is some risk as well since the ICOT researchers are exposed to other ideas and opportunities. As an intellectually curious crowd, they can be tempted by opportunities to work on technically interesting problems. For example, a Toshiba researcher, upon completion of his term at ICOT, chose not to return to Toshiba but to join the IBM Japan staff.

Estimates of the ICOT budget vary considerably depending on the source, ranging from a few hundred million dollars to as high as $500 million over the ten year period that commenced in 1981. Regardless of the actual original forecasts, however, there were government funding restrictions in 1986 and the funding matched the 1985 funding levels of about $40 million. Still, it is clear the estimate of funds allocated to ICOT should be viewed as conservative: For example, they represent only direct MITI cash expenditures; they do not include matching funds, indirect subsidies that may take the form of partial absorption of overhead costs, salaries of "loaned" workers, salary differentials between ICOT and participating companies, or donated equipment that is paid for by the companies.

ICOT certainly has a charter to perform interesting work. The Fifth Generation Computer Program, housed at ICOT, has captured the imagination of the world. The public is enchanted with the idea of new computers that can be applied to a wide range of knowledge engineering problems, including expert systems and natural language understanding. And, this project was the first, and perhaps the most ambitious, of the worldwide programs in the Technology War.

The name itself implies a radical restructuring of the technology base. The *first generation* of computers, for example, were those made from vacuum tubes; their era dates from approximately 1945 to the late 1950s. The *second generation* replaced the vacuum tubes with ferrite electromagnetic cores and discrete transistors in the years between the mid-1950s and the mid-1960s. In the *third generation* these discrete transistors were replaced with low density microelectronic circuits, and in the *fourth generation*, about 1975, increasingly dense VLSI circuits were incorporated into computers. But the *fifth generation* as defined by the Japanese is organized around different principles. It is based on the organization and architecture of the hardware, and the concept of special purpose artificial intelligence and database processors and software for faster execution of new applications.

Just as with the United States and European programs, considerable effort has been invested in intelligent VLSI CAD systems. Less well publicized has been an ICOT goal to develop research management skills in the computer sciences. The Japanese, for example, have not heretofore placed much emphasis on basic research.[14] As a result, they have not developed management skills for conducting research in this area. In our discussions with Dr. Kazuhiro Fuchi, the director of ICOT, it was clear that developing research managers was a major goal of ICOT and MITI.

One of the interesting hardware accomplishments at ICOT has been the design and development of a Personal Sequential Inference (PSI) machine intended for single-user support of program development in PROLOG, a logic programming language. PSI is now available for sale by the Mitsubishi Electric Company.

It is also interesting to examine the choice of PROLOG. At the time the Japanese selected PROLOG, its popularity in the United States was not pronounced. By selecting that language, the Japanese sent a loud and clear message that they were placing great emphasis on an independent path to their research objectives. LISP, the alternate language embraced by most AI programmers in the United States, does differ in some ways from PROLOG, but we cannot easily demonstrate significant advantages or disadvantages of one over the other. A nation must have good reasons to reject a programming language that is familiar to the bulk of the programmers working in AI-related problems. Independence is that reason in Japan.

The ICOT project has received extraordinary coverage in the Western press. Books have been written about it and an impression has been created that the West has something to fear from ICOT.[20] We don't think so. The accomplishments to date have been narrow, and major advancements in AI will not be forthcoming until the later stages of the project, if ever.

Dr. Fuchi was candid when we asked him about the role of ICOT. He said that he and Professor Tohru Moto-Oka, another cofounder of the Fifth Generation Computer Program, had desired to establish an AI research capability in Japan. At that time, there were few trained Japanese researchers with experience who were available for work on AI projects. Further, without a project underway, it was impossible to train people in AI. Their solution, he said, was to create ICOT and attract help from Japanese and foreign companies. Thus, ICOT is a creative start-up operation intended to train both researchers and managers.

NSF has recently instituted a cooperative program with ICOT that will allow Western visitors to work at ICOT. Presumably, this is to help transfer technology and research results from ICOT to the West. Although ICOT is very open to the West (for example, it promptly reports general results in the English language literature), details of its research are much harder to obtain. Thus, the new NSF program should be mutually beneficial to both ICOT and the United States.

Other National Programs

In addition to the Fifth Generation project, there are other interesting MITI and other Japanese government programs that relate to the information technology base. In some cases, they are operated through a joint research facility; in others, the research is conducted in company laboratories and only coordinated by MITI. The following discussion describes the programs most directly related to information technology.

The National Superspeed Computer Project, also known as the High Speed Computer System for Scientific and Technological Uses is a nine year program established to create a machine at least three orders of magnitude (1,000 times) faster than the fastest computer available at the time.[14] Two subprograms are underway. The first is the development of new types of high-speed logic and memory elements. The goal is to find alternatives to silicon-based technology that offer improved performance in speed with less heat. This is particularly important in view of current semiconductor packaging limitations. The second subprogram is devoted to concurrency and involves the development of parallel processing systems incorporating arrays of, perhaps, as many as 1,000 processor elements. The goal here is to find a method for combining a large number of relatively slow speed elements to achieve very fast speeds. (This problem is also receiving considerable attention in American universities.)

The participating companies are conducting research for MITI in their corporate laboratories. Fujitsu is currently working on fast and low-powered semiconductor devices. Hitachi is researching similar devices and parallel architectures. Mitsubishi is working on high-speed logic, gallium arsenide, and an experimental parallel processing system. NEC is studying the semiconductor materials and arrays of parallel processors. Oki Electric is working on materials, devices, and an experimental data flow architecture. Toshiba has a parallel processing experimental system, which includes a new parallel operating system, and parallel processing language.

This litany of each company's contribution to the MITI program is important because it illustrates how MITI and the Japanese companies in the program work together on this and other national programs to segment the research problems. Such partitioning reduces the research load for each company, minimizes duplication, exploits their particular strengths, and still enables them to share in the general results under the MITI umbrella. Apparently such cooperation in the precompetitive phase does not reduce the subsequent fierce competition between these companies that takes place in the actual marketplace.

The National Superspeed Computer Project is supported by MITI with a research budget estimated at about $100 million. MITI's ETL also participates in the project, which means that the results can be expected to be disseminated to a larger audience. Except for Mitsubishi, each of the private companies participating has its own in-house supercomputer effort separate from the MITI-supported project. Professor Hideo Aiso of Keio University serves as the manager of the supercomputer program for MITI. He replaced Dr. Tohru Moto-Oka, who passed away in 1986, as the manager of the Fifth Generation project. This has spurred speculation that the two programs ultimately will be merged.

The Next Generation Industries Program is a $472 million ten year project. Its purpose is to nurture the "revolutionary base" technologies, one of which is New Semiconductor Function elements. It has not received much attention in the West, perhaps because it is perceived as a materials research activity. Nevertheless, it has an important research subprogram in semiconductors, heretofore funded at $114 million. The research objectives are to increase greatly the number of functions that can be combined on a single chip and to study three-dimensional geometries for microelectronics.

The Sigma Program was established by MITI in 1985 as a system for industrializing and standardizing the production of software. The aims of the program are to broadly increase the awareness of the software development problem in Japan and establish uniform methods for developing software in Japan.

As defined by the Information Technology Promotion Agency of MITI, the Sigma Program goals are to install uniform software engineering standards, improve the quality and productivity of software, prevent duplicate effort, exploit improved facilities, and improve the training of programmers.

Sigma users are intended to be those people engaged in software development across Japan. The network that provides this integrated software development environment is expected to have approximately 10,000 nodes and to be constructed by the end of 1989 at an overall construction cost of $155 million. Success in Sigma would offer Japan the leverage of a standardized software environment on a national scale.

We have described earlier how the demands of the Japanese software market consume vast programming resources and create tremendous incentives to improve the productivity of its programmers. In Japan, as elsewhere, software engineering is the area in which the greatest gains in productivity can be realized. The Sigma Program, therefore, is very important to Japan's position in the software marketplace. Indeed, it is intended to address the estimated shortfall of some 600,000 programmers in the 1990s. It is a national attempt to strengthen Japan's weakest link in the Technology War.

The Optoelectronics Joint Research Laboratory is the result of the continuing quest in Japan for higher transmission speeds. In 1981, MITI started this $75 million eight year program to develop integrated optical circuits (optical switches, modulators, and such). Many Japanese companies participate in this effort. Research is conducted in a joint laboratory à la ICOT, with similar staffing arrangements. JTECH reported this laboratory to be one of the best equipped in the world.[93]

The key to success of the program will be the development of optoelectronic ICs in which both optical and electronic elements are integrated. This requires a new technology of integration with improved quality control methods. The Japanese estimate that with this technology the size of their optoelectronics industry will experience substantial growth, from $10 billion in 1990, to $50 billion by the next century.[93]

The automatic translation of languages is an important application in Japan, and it has received considerable research attention over the years. In this category, the single most striking effort in Japan is the Mu project funded by the Science and Technology Agency.[94] By the mid-1980s, eleven different companies already were producing systems demonstrating limited Japanese-to-English translation of scientific abstracts, each using the same technology based on the Mu program. This is a very impressive program and one that does not have a counterpart in the West. The program is of great importance to the Japanese since approximately $2.5 billion is spent in Japan on business and industrial translation (the worldwide market might be as high as $10 billion per year).[14]

The Industrial Giants

The Japanese companies are known for their strength, independence, and ability to work together in support of their national programs. In computing, some companies participate in every national program. They serve as the restless, but largely obedient, troops of the grand marshall, MITI.

Fujitsu, an electronics and information processing company, dominates the information processing marketplace in Japan with about a 30 percent marketshare and sales approaching $10 billion. Of this, about $7 billion comes from a mix of computer and data processing equipment. Fujitsu spends about 9 percent of its revenues on research and development: About $60 million is allocated to a staff of 700 technical employees working in Atsugi and Kawasaki, the sites of its most basic research activities. Consistent with the worldwide pattern, Fujitsu is investing substantial research and development resources in computer architecture, AI, pattern recognition, basic software engineering, knowledge engineering, machine translation, image processing, and data-flow architectures.

Fujitsu's sales first outpaced IBM's sales in Japan in 1980. Nevertheless, Fujitsu would be still smaller than most American hardware vendors if it were not for the recent appreciation of the yen. Further, with most of its revenue dependent on computers (as opposed to other Japanese computer companies such as Hitachi), Fujitsu *has* to be aggressive in the marketplace. This need to be competitive, along with a need to overcome software market differences between Japan and the United States, may explain Fujitsu's inclination to form joint ventures with companies such as Amdahl and TRW in computing, Fairchild in microelectronics, and GTE Corporation for the manufacture and marketing of telephone switchboards.

Fujitsu's name is a derivative of the Japanese word for *communications*. Organized in 1935, the company still maintains an important position in communications and electronics, such as radar. The company has a strong domestic presence (for example, only about 20 percent of its semiconductor production is destined for overseas markets).

Fujitsu accepted MITI's administrative guidance and developed M-series mainframes, a family of computers compatible with the IBM 370 architecture. In addition, its vector processor (supercomputer) series also executes IBM 370 instructions. As a result, Fujitsu serves as one of the few full product line competitors to IBM in the world.

NEC (formerly the Nippon Electric Company) is a $10 billion Japanese computer and communications company with about a 16 percent marketshare in mainframes in Japan. NEC has served as a supplier to Honeywell and Bull and, in 1986, established a joint venture with them in which it became the principal manufacturer of their large mainframes. NEC is one of the few non-IBM-compatible computer manufacturers in the world that competes with IBM across the entire product line.

As a leading supplier to Nippon Telephone, NEC is a leader in integrated communications, especially speech recognition, speech synthesis, man-machine interfaces, and distributed processing. The very nature of its product line dictates a serious research investment, and approximately 5,000 researchers work for NEC at six major laboratories. It has institutionalized its software engineering techniques for the purposes of improving productivity and quality. Its Software Product Engineering Laboratory, for example, manages software quality control for over 9,000 company programmers. NEC has an extremely young and innovative software development group with a rich programming environment and software development tools. NEC matches the more innovative American companies that are exploiting new tools, such as UNIX, special-purpose graphics, and DEC VAX computers (a typical research configuration in the United States).

Hitachi is a major Japanese computer manufacturer and vendor. Its marketshare is about equal to the IBM marketshare for mainframes in Japan. But computers only account for about 10 percent of Hitachi's revenues; its total computer revenues in 1985 were about $3 billion.[95] Hitachi spends about 7 percent of its total revenues on research and development, with approximately half devoted to basic research and half to applied research.[9] Hitachi's staff of 5,000 is spread among seven research laboratories. Research in information technology focuses on human interfaces, image processing, voice recognition, computer architecture and system evaluation, database management, computer communications networks, natural language processing, and knowledge engineering.

Hitachi has had a painful experience over intellectual property rights in the Technology War. In 1983, some of the company's employees contacted a Stanford professor and expressed interest in some of the proprietary hardware and software details of IBM's then new 3081 computer systems. This inquiry was reported to IBM, which contacted the FBI. Subsequently, an FBI "sting operation" was established, in which Hitachi employees were compromised in their attempt to obtain these technical details. The resulting arrests provoked only a mild reaction in the United States but they resulted in sensational press coverage in Japan. The IBM building in the Roppongi district of Tokyo was the scene of angry and hostile demonstrations. Many Japanese felt that Hitachi had been entrapped in a devious IBM and (U.S.) government plot. Since then, tempers have cooled. Hitachi and IBM have reached a mutually acceptable working relationship (after a period in which they operated under a court-imposed settlement).

The experience left Hitachi very apprehensive. It now is resistant to giving technical information to foreigners and is cautious in its dealings with all outsiders. The contrast in openness with other Japanese companies, is striking. Apparently, Hitachi hopes that closing down external communications will aid the company in reducing its international exposure to future allegations of misconduct and in reconstructing its image at home.

Toshiba Corporation is the tenth largest electronics company in the world. And, although it is not primarily an information processing company, it is the fifth largest manufacturer of semiconductors, with revenues of $20 billion. Toshiba spends over

6 percent of its revenue on research and development: About $120 million is spent at the Toshiba Research and Development Center in Kawasaki. Research is conducted by 1,500 people and is organized in several laboratories that are researching semiconductor materials, process technology (involved in fabrication of integrated circuits), electroceramics (an important packaging technology), CAD/CAM, and other software-related topics. The company is noted for its experimental and innovative work in software engineering. Toshiba is internationally reknown and acknowledged as a leader for its software factory practices that produce very reliable software.[96] For example, its flagship application was the development of the complex software to control a Japanese nuclear plant; such facilities demand the utmost in reliability and dependability.

Mitsubishi Electric Company is a member of the MITI and ICOT programs. The company's most important software engineering activities are at the Information Systems and Electronics Development Laboratory (ISEDL) in Kamakura, just south of Tokyo. ISEDL, which is one of eleven research and development laboratories reporting directly to Mitsubishi headquarters, centers its research on software engineering tools and, especially, metrics for measuring software complexity. Mitsubishi's ISEDL has grown phenomenally, from a staff of 150 in 1981 to over 1,800 employees in 1985. As the major software engineering center, the entire Mitsubishi company relies on it for support.

Mitsubishi also manufactures the central processor component of the PSI machine, a basic building block in the fifth generation research programs in Japan. Perhaps thirty machines have already been delivered to ICOT, participating companies, and other research facilities. These machines were widely heralded several years ago but suffer today from a growing disenchantment. For example, commercially available American workstations already have demonstrated superior performance in some applications and experiments. It remains to be seen whether the PSI 2, a VLSI version, will offer any competitive advantages over commercially available machines.

Nippon Telephone and Telegraph

The Nippon Telephone and Telegraph Company (NTT) is one of the most powerful information technology companies in the world. NTT was modeled after AT&T and its European counterparts and offers excellent communications services to the second largest economy in the world.

The Japanese government recently deregulated NTT, and it is now beginning to experience the same chaos among its suppliers and customers suffered by AT&T. NTT's research laboratories, however, have not suffered the same fate as AT&T's Bell Laboratories. NTT research continues to play a principal role in the country's activities through the Electro-Communication Laboratory, which functions close to that of a government research laboratory. With much of ECL's capabilities similar to Bell Laboratories, its central role in most MITI projects provides a powerful boost to the country's programs. Since senior ECL officials serve on many consortia and as managers in MITI laboratories, Japanese industrial concerns receive information about NTT's research results almost directly.

As explained in Chapter 2, the Japanese and Chinese alphabets are awkward to use with computers. Thus, there is an incentive for the Japanese (and Chinese) to improve their man-machine interfaces and communications technologies that deal with these alphabets. Modern digital communications technology offers that opportunity, and ECL is active in this area. In fact, NTT is currently establishing an integrated digital network system (INS) that will rely heavily on fiber optic technology, VLSI, and communications satellites. When completed, NTT hopes that INS will effectively integrate all of the existing telephone, telex, facsimile, and data networks in Japan. NTT expects to spend $86 billion to complete the system by the year 2000. Its commitment to INS is expected to infuse from $8–13 billion in the next ten to fifteen years into the Japanese optoelectronics industry alone. Such an injection of support can be expected to be accompanied by substantial advancement in the technology.

The INS participating companies are expected to use the various media to offer a wide range of consumer services such as home banking, shopping, travel information, plane and train schedules and ticketing, movie and theater schedules and reservations, financial information, health information, news telephone service for the hearing impaired, popular book and record reviews, and home-study courses. Similar programs have been tried in the United States on a much smaller scale but have not been very successful. France has also experimented successfully with this technology. The Japanese, who are noted for their fascination with gadgets and their natural curiosity in high technology, can be expected to be more enthusiastic users of these services.

In other areas, NTT routinely spends $300–350 million annually for advanced research and development in computers, telecommunications, and semiconductors and ECL is charged with managing most of that research. Work is organized into four laboratories: Musashino, which specializes in integrated networks, switching, and communications processing systems, as well as general fundamental research; Yokusuka (data communications systems); Ibaraki (materials); and Atsugi, whose emphasis is on microelectronics research such as VLSI, CAD, gallium arsenide, and packaging.

Doing Business in Japan

Much has been said about the difficulties of doing business in Japan. Americans complain vociferously about Japanese resistance to their products. Indeed, competing for marketshare in Japan is difficult for foreign companies: It is subject to all the Japanese idiosyncrasies about buying domestic products, quality and reliability, and so forth. And Japanese products are difficult to compete with and are usually protected by tariffs and other trade barriers, such as quotas. Nevertheless, some American and other Western companies have been successful; IBM Japan, Kodak, Xerox, Coca-Cola, Mercedes, and Rolex are only some of the examples.

IBM Japan has been successful both in penetrating the market and in attracting and retaining Japanese staff (in one case even recruiting them from that bastion of Japanese strength, MITI's ICOT laboratory). Still, it is difficult to explain this singular success. Any attempt to classify IBM Japan leads to contradiction. It is considered a Japanese company by many Japanese, but a foreign company by others. Although it has been able to achieve a 30 percent marketshare (in head to head competition with Fujitsu, NEC, and Hitachi), it has not gained the dominance it enjoys in the West. It remains to be seen whether its 10,000 Japanese employees will make it possible for IBM Japan to improve its half-in, half-out position in the Japanese information technology society.

Another way to succeed in Japan is by following the model set by Xerox. By forming a joint venture with Fuji Film, each partner brought its specialty to the table. Xerox exploited its expertise in copiers, thin film, and microelectronics, and Fuji brought film, color, chemistry, and manufacturing strengths. Both have been successful in both the Japanese and U.S. markets.

One can do business in Japan. The Japanese markets are not closed to outsiders and can be penetrated by companies with quality products and a readiness to embrace local sales practices. In the 1920s, General Motors did have 90 percent of the automobile market in Japan; Japan Air Lines is a large user of Boeing aircraft in the 1980s.

Summary

There is no question that Japanese companies are making an enormous investment in an attempt to gain the prominent position in information technology for Japan. The sum of research and development funds among the companies discussed in this chapter is about $3 billion, but that is equivalent to a much larger figure in the United States since the Japanese work only on problems that have received the consensus stamp of approval. This Japanese investment must be taken very seriously in the Technology War. Still, it is not necessary to concede defeat just yet. The Japanese face a host of serious problems: They are totally dependent on external sources for their energy, and their food supply is controlled by a powerful political syndicate that keeps consumer prices high and foreign food products out of Japan. Japanese workers who have comparable incomes to American workers suffer a one-third loss in their buying power compared to the Americans; buying power for which they work longer hours with fewer rewards.

Further, the Japanese population is undergoing great change as it ages and experiences a growth in affluence. The greatly appreciating yen has had a devastating effect on their exports, and unemployment is up to 3 percent, a very high rate for the Japanese system to bear, especially when this figure does not represent all categories of workers.[7] The trade wars have intensified, and the Japanese are being undercut by the other Asian countries that find themselves in the same comparably favorable position that Japan once enjoyed.

Technically, the Japanese continue to suffer a profound disadvantage in software, and this may be at the root of their problems in the workstation and personal computer markets. They also have a difficult time with rapidly innovating companies such as Apple, Lotus, or Sun. These companies form one of the West's major strategic reserves.

Will these problems prove insurmountable in Japan? We don't know. We do know, though, that Japan will not attack these problems piecemeal. They will analyze their problems on a systematic basis, debate the choices, reach a consensus, and march forward together. Contrast this process with their other international competitors, the subject of the next chapter.

10

Other International Programs

The Technology War is a global contest between the United States, Japan, Western Europe, and the other nations of Asia. But it is not a uniform struggle everywhere; in the different regions of the world, each of the competitors has its own tale. For example, in its battle to stay competitive, the highly fragmented community of nations in Western Europe suffers from a lack of coordination and standards, a host of small marketplaces (each lacking critical mass), and an aging industrial plant. It has an entrenched and conservative leadership that has neither recognized the value of high technology nor has volunteered, in the past, to accept the risks necessary to shift the manufacturing base from the Industrial Revolution to the Information Age. The nations only recently have grudgingly accepted the need for the formation of alliances in the Technology War.

In contrast, a number of small southeast Asia countries have been quick to grasp the opportunities afforded high technology suppliers, first as manufacturers of foreign-designed and licensed equipment, and, more recently, as independent producers. However, these small countries lack marketshare and the benefits of a sophisticated technology infrastructure. In dealing with their problems they have formed alliances for some of the same reasons as the Europeans.

In our treatment of the competitors in this chapter, we focus especially on these strategic partnerships and relationships. The first of these, which merits some attention in Europe, was created specifically in response to Japan's announcement of the Fifth Generation Computer Program. This first continent-wide reaction is managed by the European Economic Community.

ESPRIT and Other EEC Programs

ESPRIT, the European Strategic Program for Research in Information Technology, is a research program founded in 1982 and managed by the EEC. Half of ESPRIT's funding is expected to come from corporate participants; the remainder is provided by the EEC.

The $1.4 billion program is directed exclusively at problems in information technology.[97] The Commission of the EEC justified a program of this scale on the grounds that any European response had to be comparable with the programs of its competitors, that is, America and Japan. Nevertheless, the program is billed as pro-Europe rather than anti-American or anti-Japanese.

Adopting a Japanese phrase, ESPRIT has targeted technology still in the "precompetitive" phase. The program originally focused on five technical areas: advanced microelectronics, advanced information processing, software technology, office automation, and computer-integrated manufacturing.

In 1984, ESPRIT completed a one year pilot phase supporting about three dozen projects. The following year ESPRIT selected ninety-five projects to fund. The two year total (1983-1985) was over 170 projects, involving over 450 industrial firms and 100 universities. About 150 of the firms were in the small to medium class. As of 1987, the program supported approximately 3,000 researchers.

In our discussions with EEC officials, the goals of ESPRIT were elaborated with varying degrees of idealism. Such variations exist because, at best, the program can foster genuine European unity and provide a critical mass for investigating new areas of technology. On the other hand, there are cynics who suggest that ESPRIT, for all of its lofty goals, really amounts to nothing more than a subsidy from EEC funds to the twelve largest European companies. Regardless, we consider ESPRIT woefully inadequate to meet the needs of Europe in the Technology War. It is a small program, for example, representing only ten times the funding of a single American research consortium, MCC.

Although participation was originally limited to EEC countries, there has been some recent movement to broaden the research base in ESPRIT to include six non-EEC nations: Austria, Finland, Iceland, Norway, Sweden, and Switzerland. It will be interesting to see if Jefferson's "nations of eternal war" can cooperate effectively in ESPRIT.

A second program, ESPRIT II, is now in the planning phase. Hoping to start in 1987, ESPRIT II calls for a tripling of effort and the application of 30,000 man years to microelectronics, applications, standards, and information processing systems. This development of standards is particularly important because of the fragmented nature of the European market. Not only can the enforcement standards make the European marketplace a single larger and more coherent entity, standards also can force the Americans and Japanese to comply with them if they wish to sell products in Europe.

Despite these lofty plans, ESPRIT has been forced to consolidate some of their programs. Research goals have not been met, and some retrenchment is necessary. The computer-integrated manufacturing and office automation programs have been consolidated as have the advanced information processing and software engineering programs. Nevertheless, ESPRIT has produced a great level of continental scientific cooperation and coordination. It has resulted in technical progress in VLSI topics, and the formation of manufacturing coalitions. The question is whether these results will sustain themselves and whether the governments of Europe will continue to work together in support of the goals of ESPRIT. We are skeptical that such cooperation will continue indefinitely.

The EEC's executive commission also is planning a massive ten year drive to improve Europe's tangled telecommunications system. They hope to facilitate the development of an integrated system that uses optical fiber cables, satellites, and digital switches. Such a system could ultimately cost European governments and industry more than \$200 billion.

The RACE program, Research in Advanced Communications technologies for Europe, is Europe's first attempt in the Information Age to rationalize its loosely coupled federation of Post, Telecommunications and Telegraph companies (PTTs) and to develop a coordinated response to the Japanese challenge posed by Nippon Telephone and Telegraph's Information Network System.

The first step is a \$29 million pilot research project. The objective of this preliminary research is to develop a consensus on the specifications of a future Pan-European broadband system. RACE, operating in an ESPRIT-like mode, would then share half the research costs of companies and research institutes that research problems in related communications topics.

ECRC and Eureka

The large companies of Europe and their corresponding governments occasionally do implement joint programs in areas in which they find common need. Although Europe is often slow to reach consensus, two programs in the Technology War stand out. These are the European Computer Industry Research Center (ECRC) and the Eureka activities.

In January 1984, Bull, International Computers Limited, and Siemens agreed to form a joint venture, the European Computer Industry Research Center, based in Munich. The objective was to facilitate collaboration in long-range research in artificial intelligence and expert systems. ECRC probably was also established as a technical hedge against the possible failure of ESPRIT.

The ECRC grew modestly to a staff of 50 by the end of 1986. The average yearly budget is $7.5 million. The three sponsors share all costs in the laboratory fully. Similar to ICOT, the laboratory members are on special leave from their companies. The Director of ECRC is Dr. Hervé Gallaire, who received his graduate education in the United States and has had a distinguished research career in French laboratories and industry. His appointment to this position is an indication that the sponsors take ECRC's research goals very seriously.

ECRC is organized into four main groups, studying logic programming, knowledge-based systems, symbolic computer architectures, and man-machine interaction. All of the projects are closely related in contrast to the distinct compartmentalization of programs at MCC.

On the important and complex issue of licensing, the sponsoring companies get free licenses to products and systems developed at the laboratory. Staff rotate in and out of the laboratory à la ICOT, facilitating technology transfer as well.

Another European attempt to keep pace with the United States in research is the Eureka program. The name is a loose abbreviation for European Research Cooperation Agency. One of its early objectives was to staunch the increasingly painful brain drain that was feared would result from the massive SDI funding in the United States: The Europeans feared that American research funds would attract a growing number of their scientists to the United States.

In addition, Europeans have been reluctant to participate in SDI. Its military focus has been politically undesirable and created a host of problems for the European nations. As a result, it has not been clear whether the nations of Europe would ignore SDI, cooperate with it, or compete with it.

Eureka offered a more desirable alternative to Europe. As a civilian program with, presumably, military applications in the background, it was far more acceptable to the public. It also served to appease the Soviets, who have been vociferous opponents of SDI.

Regardless, the history of this program tells a lot about the prospects for European cooperation in information technology. In 1985, NATO countries and France began to plan a possible joint research program that would serve as a counterweight to SDI. Six initial problem areas, similar to the SDI objectives, were singled out for attention: optoelectronics, new materials, high powered lasers and particle beams, AI, and high-speed microelectronics. Catchy new names were proposed for the projects such as Eurobot for a communal robot and Eurocom for a European communications system.

The first responses to the plan were enthusiastic. France proposed an initial $160 million and the Germans considered an investment of $80 million. In the summer of 1985, the Americans invited the European countries to participate directly in SDI research and projects, offering them about $30 million. But Germany offered $25 million to Eureka instead.

By the end of 1985, the plan began to unravel. Germany backed away from its financial commitment. France clarified its intention to use old or previously allocated funds that would only be available by cutting other French programs. Curiously, Britain, which had been lukewarm all along, began to be more interested. But by 1986, Eureka still had no permanent home and a most uncertain future. The Soviet Union was raising political questions as to whether Eureka violated the Helsinki Agreement. Bulgaria, Hungary, Czechoslovakia, and the German Democratic Republic were expressing interest in joining Eureka. (That suggestion raises interesting specters for the Coordinating Committee, the group of Western allies and Japan that monitors high technology exports to the communist bloc.)

Still, Eureka may be working. Over forty proposals have been submitted by 1987 and $200 million is earmarked for twenty-five projects. The projects include the development of an X-ray lithography system, the establishment of a custom chip company, and a small supercomputer system. Nevertheless, for as long as this process has taken, we believe this tale justifies the current round of Europessimism. There are great plans delineated, a plethora of political posturing, and then little actual accomplishment. Popular American magazines use words like Eurosclerosis to describe this sort of condition. Our sad tale demonstrates only too well how politics and bureaucracy have delayed the development of critical technology in Europe.

In addition, we believe the European goals are too limited. Consider, for example, the limited computational objectives of the ESPRIT program as described by the EEC:

> The achievement of technological parity with, if not superiority over, world competitors within ten years.

Parity is an inadequate goal; playing catch-up is unlikely to motivate many technologists.

Western European National Activities

There are several related research and development efforts in information technology in some European countries. These programs collaborate with ESPRIT on many research topics. In this regard, the Alvey program in the United Kingdom, and several efforts in France, deserve special attention.

Great Britain

After the original fifth generation announcement, the British academicians rushed to establish a committee to consider the implications to British competitiveness of success in the Japanese program. The committee recommended a restructuring of information technology research in the United Kingdom. This resulted in the Alvey report, which was released in 1982. It was widely circulated and won unanimous support from industry. The committee originally recommended that 100 percent of the research costs be supported by government funding but, as approved, the program provides for a 60/40 percent cost sharing between government and industry. This change, unfortunately, had a negative impact on many small firms that wished to participate in the program but lacked the funds. Indeed, in the absence of readily available investment funds in the United Kingdom, only about twenty-five firms could afford to participate in the program on a significant basis.

The program is named for John Alvey, director of research for British Telecom, who chaired the study committee.[63] It was established under the management of the Department of Trade and Industry, and its activities are coordinated with the British Ministries of Defense, Education, and Science. The program is expected to cost from $550 million to $580 million over the course of its five year life.

During the planning phase of the program, the Alvey Committee used the MITI and ICOT programs and DARPA's Strategic Computing Program as models. Just as in these programs, the Alvey Program's research emphasis is on software engineering, VLSI, man-machine interfaces, and intelligent knowledge-based systems. The program certainly serves as Great Britain's grand marshall when viewed in the context of the other European programs, although its size is modest in comparison to the American and Japanese programs.

Alvey research in microelectronics is intended not only to enhance the speed of integrated circuits generally but to develop special-purpose chips for knowledge systems applications. Research is also being conducted in gallium arsenide and packaging topics. The VLSI research focuses on functional-logical and data-flow machines as well as smart databases. The British Ministry of Defense's Very High Performance Integrated Circuit Program (already in existence) was absorbed into the VLSI section of the Alvey project. This relationship to defense research is similar to the pattern in the American programs.

The British do feel threatened by their technological shortcomings. After the Alvey program was established, a senior British official made a point of telling us the following (paraphrased here):

> The British take the Japanese challenge in computing very seriously. As a matter of national policy,

Britain cannot be left behind. Its industrial future depends too heavily on computing. Therefore, the British are looking forward to international cooperation and joint ventures to improve their technological base. They want very much to find a way of doing this with the United States. Lacking such a possibility, they will do it with Japan.

So the British government understood the significance of the Technology War from the start. They organized quickly and created a well planned, if small, program. But what about British industry, particularly their largest computer firm, International Computers Limited (ICL)?

ICL is the principal British computer firm, with a sales volume of about $1.5 billion. In the summer of 1984, it was acquired by Standard Telephone and Cables, a former ITT company. As is the case with most European suppliers, ICL enjoys a special supplier relationship with its government.

ICL is quite active in the United Kingdom and NATO research establishments and participates in the Alvey and ESPRIT programs as well as in ECRC. It is active in CAD, having acquired a government-owned research and development center in Cambridge. Nevertheless, the company has failed to establish a strong presence in the international market. We wonder how it would fare without its preferred relationship with the British government, which guarantees it a subsidized and indefinite existence. ICL serves as an example of the perils of direct government support of a high technology company; the lack of competition leads to little technological innovation and the ultimate erosion of international marketshare.

Another British firm of some importance is General Electric Company (GEC). It is a $7.5 billion firm with considerable strength in the new technologies. Through the Marconi companies, it is very active in electronic systems and components for defense, energy, and computing applications. In 1984, GEC's Telecommunication and Business Systems Division alone exceeded $1.1 billion in sales of modern digital switching equipment. In research, GEC places some effort in VLSI, gallium arsenide, and computer science. And it too is looking to AI to help deal with its software engineering problems. Nevertheless, in most computer science circles, GEC is considered overly conservative and it is not known for great innovation.

Other important companies in Great Britain include Plessey, an electronics firm active in military communications, and Ferranti, a computer and electronics firm. It is also worth noting that INMOS (a subsidiary of Thorn EMI) was established to help the United Kingdom expand its presence in the microelectronics marketplace. Several hundred million pounds were invested in INMOS. But the company operates a microchip plant in the United States, in part because there have not been enough high-quality production techniques in the United Kingdom to meet the fabrication requirements of the next generation of chips. Sadly, economics have forced INMOS to abandon the RAM marketplace and to reduce its work force in the United States (by two-thirds in 1986). Of course, American memory chip manufacturers have experienced similar problems.

France

French activity in information technology systems is so centralized that the government is involved in almost every aspect of the computing field. Centralization began in 1982 when the Mitterrand government nationalized key companies in the electronics and computer fields. By 1987, France had the largest government-owned data processing sector outside of the communist countries. It has been estimated that as much as 28 percent of French goods and services are produced by companies in this sector.[22] In fact, ten of the largest fifteen French companies, as well as almost all the banks, are owned by the government. In this case, government ownership has not been all bad; the major high technology companies improved their financial position with the aid of some $5 billion of government funds.

After five years under this organizational structure, France's new premier, Jacques Chirac, announced plans to "privatize" between forty and eighty information processing companies. Using Gallic linguistic precision, a distinction is drawn between *denationalization* (immediate sale of shares in the government-owned companies) and *privatization* (a gradual sale of stock over five years). This will be a $21 billion sale if it actually materializes. But it is not clear whether it will; for example, one highly placed observer has stated, "The French are afraid that they will lose their Bulls to IBM," by which is implied the loss of the vitality of their flagship high technology company.[98] In any event, it seems clear these high technology companies will experience a change in their international competitive posture. That probably will not be the case, however, with the French PTT, which, despite the deregulation of AT&T, NTT, and British Telecom, seems destined to preserve its government monopolies.[22]

One of France's general national programs, referred to as the Groupe de Recherche Coordonné, is made up of cooperative units under the management of the Centre National de Recherche Scientifique. Participants come from universities, national laboratories, and companies such as Bull, Compagnie Generale de L'Électricité, and Thompson-Brandt. The principal research topics are man-machine interface (natural languages, speech synthesis, recognition, and vision), robotics, relational database organization, intelligent knowledge-based systems, and AI software engineering. The main funding sources include the French Ministry of Industry and Research, French Telecom (Centre National d'Études de Télécommunication), and the French Ministry of Defense (Direction de Recherche d'Études et Technique, the French counterpart of DARPA in the United States).

In a separate development, the French government plans to install a nationwide cable television system using lightwave technology. The first order was placed in spring 1984 for a $170 million hybrid coaxial cable, fiber optic system to wire over 300,000 homes; 6 million homes are to be wired by the mid-1990s. This is another substantial competitive response to the INS program.

In software, the French have been quite active. In fact, it was a Frenchman, Alain Colmerauer, who conducted the fundamental research on the logic programming language PROLOG. Moreover, another Frenchman, Jean Ichbiah, performed the key language design work on the language Ada (discussed in Chapter 7).

In other related high technology fields, France also has significant strength, such as in nuclear technology, aviation, and telecommunications. Its problem is not unlike that of the United States. There is no consistent, coordinated French program in information technology. When the Mitterrand regime took office, research and development spending did rise for a period of time. But the current government has reduced research and development spending by 8 percent and folded the Ministry for Science into the Ministry of National Education. And the new French government, ever anxious to use French genius, has just appointed C. de Vignemont as a ministerial advisor for high technology.[95] Monsieur de Vignemont is fifteen years old!

The French industrial sector has some important members. A strong high technology company, for example, is Thomson-CSF. In 1985, it invested 18 percent of its sales, about $800 million in research. Thomson's basic business is defense electronics and communications, with worldwide sales of over $4 billion. It is the key semiconductor manufacturer in France and a powerful player in the Technology War.

However, the leading French computer and data processing company is Bull. It is the jewel in the French crown as ICL is in Great Britain and Siemens in West Germany. Bull, however, has had a troubled history, and its sporadic behavior over the last several years reflects the contradictions found in the behavior of the French government. As an old data processing company, Bull carries a reputation for being unwilling to try anything until it has been done by IBM. In one case, a new software project proposed by engineers was denounced because it was thought to be too visionary. Two years later, after IBM had announced software products based on these concepts, the same engineers were asked to develop competitive software products in two months.

Management seemed insensitive to the fact that such products take years rather than months to complete.

In spite of past weaknesses, Bull is improving. Under the leadership of Jacques Stern, customer satisfaction has risen. Bull has a new manufacturing and marketing agreement with Honeywell and NEC that should offer it new marketing opportunities. But old methods die hard. Research and development are still defined inside Bull to include maintenance. The maintenance of its old operating system, therefore, takes a major share of the research and development budget, leaving little for new technical innovation.

To summarize, Bull is a major company (ranked sixteenth in the worldwide industry).[95] It has been as successful as it is through government support, cross licensing, and a very good sales force. It is vulnerable to foreign competition, although the French marketplace has remarkable similarities to the Japanese marketplace in its consumers' passion to buy domestically produced goods. The effects of privatization on Bull and its long-term viability are unclear.

West Germany

West Germany has groups working in a wide range of related areas, including expert systems, natural language, robot vision, man-machine interface, programming environment, and machine architecture. Expert systems projects are underway at several universities as well as at private firms, including Nixdorf, and Siemens. In addition, a federal program in expert systems technology is supported by the Bundesministerium für Forschung und Technologie (BMFT), the government's research ministry. West Germany also has a national project for the development of parallel processing systems.

The government of West Germany and its industry have also begun a major experimental communications project to follow the Japanese and the French, labeled the Broadband Integrated Fiber Optic Network. It represents the first step in the German Bundespost's plans to replace its copper wire systems, over the next three decades, with an optical fiber-based integrated telecommunications network. Over $70 million have been spent on the project. Another system is being installed to support an intercity broadband network for videoconferencing and high-speed text and data transmission. The first stage will link the fourteen largest West German cities, probably with about 1,000 subscribers. By 1988 or 1989, about forty to fifty cities and towns will be connected via the network.

In West Germany, the Technology War is not restricted to its more conventional forms. An increasing number of terrorist attacks there are directed against technology. The July 9, 1986, assassination of nuclear physicist Karl Heinz Beckurts, Siemens' research director and liaison to the American SDI, is one example; the July 25, 1986, attack on Dornier (working with Sperry Univac and Bonn on Star Wars' programs) is another. These attacks are likely to proliferate as radical groups become more familiar with the political impact of high technology.

Still, these acts against Siemens may reflect the fact that Siemens is the key West German electronics and communications company in the Technology War. It had sales in 1984 of about $16 billion. In 1985, it was the number two supplier of data processing equipment (behind IBM) in Europe. Siemens is one of West Germany's most prestigious high technology firms, employing over 30,000 people in research and development. With strong support from the BMFT, it also is active in nuclear energy and materials. Its recent joint venture with GTE demonstrates its commitment to communications: Paying GTE over $400 million, Siemens will absorb GTE's transmission units in Italy, Belgium, and Taiwan.

Siemens is pursuing most of the fifth generation technologies. The company is a founder of the ECRC and is quite active in ESPRIT. It places special emphasis on large architecture systems, personal computers, and VLSI. In this area, Siemens established an investment program of about $300 million in 1984, with a target of producing a 1 megabit RAM chip by 1987. Its 1989 goal is a 4 megabit chip. Siemens is also working on fiber optics, workstation processors, and integrated digital networks.

Servicing over 1.5 million customers, Siemens is, in our opinion, one of the most powerful information technology companies in Europe. Sadly, its computing equipment does not seem to command the marketshare that the Japanese or American computer manufacturers enjoy; nor does it exude the calm competence of Nixdorf, a much smaller but more commercially focused West German data processing firm. Nevertheless, we look to Siemens to provide most of the scientific leadership in West Germany in the Technology War.

Other European Companies

There are other European programs and strong industrial firms whose capabilities play an important role in the Technology War. Throughout Europe, perhaps no other firm deserves as much attention as Philips, a large electronics firm based in Holland. As a $20 billion conglomerate, it ranks in the same class with NEC and has a long tradition and excellent reputation for its work in basic research in all of the sciences.

Although not a major force in computing, Philips plays a substantive role in the European fifth generation programs. It is quite active in the microelectronics program in ESPRIT and has formed a collaborative program with Siemens aimed at the development of submicron technology.

Philips has 4,000 people engaged in research and has not restricted its base to Europe. In 1984, it established a research laboratory in the United States (as did Olivetti in California and Siemens in New Jersey).

Other European companies of some importance are Olivetti (which has major cross licensing, manufacturing, and sales agreements with AT&T), STET, Nixdorf, AEG Telefunken, and the Saab family of companies. Most of these companies have research activities that support the ESPRIT and other national programs in Europe. Nevertheless, they represent just another splintering of the continental effort since each company represents, in general, the same position in its country as ICL does in Britain.

Asian Projects

In Asia, Korea, Taiwan, Singapore, and Malaysia are growing centers of technological excellence that are offering a growing competition to Japan. They have recognized that technology is the key to rapid industrialization and market power. China, the Big Dragon, also has an appetite for high technology and represents, perhaps, the largest potential market in the world. In addition, China has an outstanding scientific tradition that will fuel its hunger for the West's technology. It will become a competitor in the future. Although it is not currently a threat in the Technology War, its very size demands attention.

People's Republic of China

The People's Republic of China (PRC) is a fascinating example of a less developed country whose problems are compounded by size. The Big Dragon, for example, has a population of over 1 billion people, but (according to the World Bank) it is also in the 136th position in the world on a per capita income basis.

There are other staggering problems, such as in education. In 1987 there were 250 million illiterates in China, with 10 million middle and high school students having no further educational opportunities because of a lack of vocational schools.

China's Cultural Revolution, which began in 1966 and lasted ten years, was a disaster of major proportions to the country. Essentially, China lost a generation of educated personnel; and teachers, professors, and skilled technicians are now in very short supply (especially those between twenty-five and forty). Furthermore, during the Revolution, many of China's leading scientists and engineers did not practice their professions, and their skills and knowledge became outdated. The "divorce of theory from practice" and the "divorce of research from production" were declared by the Red Guard to be major faults of some Chinese researchers. A great number of Chinese scientists were subsequently persecuted, labeled the "bourgeois reactionary academic authority" or "young successors to revisionism," and sent to the rice paddies.[99,100]

China is embarked now on an ambitious program to modernize its industries.[101] Its primary strategic objective is to quadruple the GNP by the year 2000. A second goal is, in twenty years time, to reach the economic level of the middle-income countries. Its third goal is to catch up with the developed countries by the year 2050. However, China has organized its research and development programs based on the Soviet model, which features centralized management. This policy has led to the establishment of 10,000 research institutions with over one million employees. It is inconceivable that China can coordinate such a large body of independent entities in any systematic manner.

Still, now in their sixth Five Year Plan, the Chinese have adopted new policies including using Western ideas such as venture capital, to accomplish their goals in basic research. A key to venture capital, however, is the underlying tax laws and the corresponding compensation offered to the investors; China has a long way to go to put these incentives in place before the venture capitalists of China reach measurable proportions.

An area that attracts the attention of the government is China's technology lag with Western nations and Japan. As part of an attempt to absorb more foreign investment and to accelerate the development of advanced technology, more than fifty laws and decrees have been enacted in the past few years, covering joint ventures, income tax, special economic zones, foreign exchange control, and so forth. The patent law, proclaimed in 1985, is one of the more important laws. It encourages the import and development of new technologies. In addition, China, in 1986, devalued its currency 16 percent relative to the U.S. dollar, which, along with its protectionist regulation, should help its domestic technology industries develop and simultaneously encourage exports.

Although it *is* a backward nation in many respects, China is determined to modernize itself, and it has taken impressive strides in the last few years. Also, China has a strong tradition in science and technology and a record of creative achievement. It has been at the forefront, for example, of the latest work on superconductivity.[79] China did invent gunpowder and the first rocket. These traditions can be expected to survive the trial of socialistic bureaucracy. The People's Republic of China will be an important player in international high technology sooner rather than later. We think China has a good chance to ultimately absorb technology in the same fashion as we have witnessed in the Japanese and Koreans.

The Little Dragons

If China is known as the Big Dragon, then South Korea, Singapore, Taiwan, and Hong Kong are known as the "Little Dragons."[102] They are powerful! Their economies have grown at an average rate of 9 percent a year for twenty years. But, competitive problems, fueled by rising labor rates, have slowed their growth and are impacting them, too. The comparative advantages continue to shift.

Each of the Little Dragons has different needs and requirements. Their market sizes vary considerably, each plays a different role vis-à-vis the PRC, and each is in a different state of industrial development. But *all* of them are in pursuit of the information technology markets.

The Republic of Korea (ROK) is clearly challenging the Japanese, for example, in many high technology markets and has publicly stated its intention to beat Japan by forming strategic alliances with the United States. This may be plausible. In twenty-five years it increased its GNP per capita by a factor of 20. One of its major competitors, Singapore, a strong state in its own right, reports that their competitive position against Korea weakened 35 percent (as measured by a mix of productivity and wages) in just six years.

Korea is formidable on many fronts, such as computers, consumer electronics, and automobiles. Indeed, in referring to the Korean automobile industry, the *New York Times* described its entry into the United States as "The Next Wave." Perhaps more impressive is the fact that Korea is also selling automobiles in Japan!

Besides Korea's strength in automobiles it is also active in industrial robots, personal computers, and microelectronics. It manufactures, for example, one of the best price-competitive alternatives to the IBM PC. And in microelectronics, it has been very successful in forming strategic alliances. To date, Korea has invested hundreds of millions of dollars in the Technology War. Samsung, for example, expects to complete an investment of $500 million in microelectronics fabrication (under a joint technology agreement with ITT) by the end of 1987. The Korean Institute of Electronics, a government-owned facility that was funded by the World Bank, is active in leading edge VLSI technologies. Gold Star, one of the largest electronics companies in Korea ($6 billion in sales), has a technology agreement with Western Electric and could invest over $150 million in microelectronics fabrication facilities. Daewoo, the owner of ZyMos, has agreements with Northern Telecom and has programmed several hundred million dollars for its investments in microelectronics fabrication. The Hyundai Group, which manufactures automobiles (sales of $9 billion), is estimated to have a $500 million investment plan for DRAM production over the next five years.

Korea now is drawing charges of dumping semiconductors. Despite the charges, we would expect the ROK to continue to be aggressive on pricing all of its electronics products. It enjoys all of the advantages that Japan had in the 1970s: a skilled workforce, cheap capital and favorable tax policies, low priced labor, and a standard of living that is relatively low compared to its neighbors. This will enable Korea to maintain very fierce price competition while it simultaneously increases the rewards to its labor force. Such a strategy, presumably, is well understood by the Koreans. We can see signs of it in the growing number of joint ventures with U.S. firms. Korea suffers from a disadvantage, however, that the Japanese never experienced. Its competitors are now very sensitive to the threats posed by its economy, and Korea must expect tariff battles (and wars over intellectual property rights) that will undoubtedly interfere with its expansion.

Singapore's 1985 and 1986 recession forced the setting of new strategic directions to enable Singaporeans to continue to enjoy the rapid increases in their standard of living that they had experienced between 1965 and 1985. Among its recommended new policies and directions is *one* fundamental technology: information technology, composed of robotics, artificial intelligence, microelectronics, laser technology and optics, and communications technology.

The goals of its information technology policy are clear:

> (a) to encourage all industries to exploit and apply new advances in technology as widely as possible; (b) to develop competence in selected new technologies whose future importance is clear, and where we [Singapore] have a comparative advantage ... *a specific strategy should ... be drawn up to exploit information technology.* [emphasis in original][103]

Singapore considers it as a necessary ingredient to the survival of its industrial base. It says

> It is not enough just to computerize. What is needed is strategic and creative exploitation of information technology by our industries.[31]

Within Singapore, the policy and strategy on computing is managed by the National Computer Board (NCB). The NCB was established in 1980 by the Defense, Telecommunications, Education, Finance, and Industrial agencies of Singapore. It places considerable emphasis on maintaining focused research and industrial programs in Singapore (to some extent like MITI in Japan). Its current target is establishing strong software capabilities in Singapore with an emphasis on knowledge-based systems,

CAD/CAM, and telecommunications. Another objective of the program is to increase the computer literacy of all Singaporeans.

Despite its investment in a variety of national programs, Singapore lags the ROK by almost a factor of 3 in its ratios of research and development expenditures to GNP and lags the United States by a factor of 6. However, Singapore's proportion of research scientists and engineers to its total work force is comparable to the ROK and only lags the United States by a factor of 3. Seeking to improve this ratio even further, Singapore's research and development strategy includes a plan to raise the social status of scientists and engineers.

We rate Singapore very high on the list of Pacific Rim competitors that can menace Japan. Sensing the threat, the Japanese International Cooperation Agency is already actively participating in joint programs in this region of southeast Asia.

Taiwan has had an aggressive program of economic development. In the last six years it improved its productivity position by about 15 percent with respect to Singapore. Indeed, its productivity growth was greater than that of any of the Little Dragons, perhaps because of its heavy dependence on technology, borne out of national security requirements.

Similar, in a strategic sense, to Israel, Taiwan has found it necessary to acquire modern technology to overcome the military advantage that accrues to its larger neighbor. As a result, the Taiwanese government has expended great efforts to promote a strong domestic computer and software industry. The country has invested over $200 million in a VLSI facility and has ambitions in the office automation market. Indeed, it has targeted for 2 percent of all the worldwide information technology markets by 1989. With approximately 100 software houses, Taiwan hopes to capture $3 billion (about 5 percent) of the worldwide software market by 1989.

Hong Kong has a long history as a mercantile outpost for mainland China. Its freewheeling economy is legendary for the opportunities it presents to entrepreneurs. However, as one of the Little Dragons of southeast Asia, it represents more of a financial and manufacturing threat than a technology threat. There is no research base of any consequence there, and it is unlikely that one will develop in the foreseeable future because of the impending shadow of PRC control that is due in 1997.

Hong Kong offers an interesting door to mainland China and could serve as a low-cost broker to the Chinese worker. It is already turning to mainland China for textiles and toys, for example, and it serves as the financier for the other Little Dragons' investments in China. But Singapore offers serious competition to Hong Kong in any relations with the PRC, and it has less controversial relations with the Chinese. Their distance, perhaps, makes them less threatening to the PRC.

In the Technology War, Hong Kong continues to suffer from a reputation for software piracy. Its products are viewed with skepticism and suspicion in the United States, and companies such as Apple are continuously monitoring Hong Kong products for possible infringements of their copyrights.

Although not in the class of a Little Dragon, Malaysia's microelectronics industry represents a large segment of the world's chip manufacturing fraternity. In fact, Malaysia is reputed to be one of the world's largest producers of integrated circuits, its output having reached about $1.67 billion in 1984 alone. The industry provides employment directly or indirectly to nearly 100,000 people and contributes a significant proportion of the nation's foreign exchange and income.

With so much skilled personnel employed in microelectronics, there exists a potential to upgrade the technological base of the industry further from a largely labor-intensive assembly operation to a more creative activity involving both circuit and system design. Addressing this potential, in 1985 the Government of Malaysia created an ambitious public funded research and development institute entitled the Microelectronics Institute of Malaysia to conduct research and development in the field of microelectronics and to encourage and support the creation of new industries based on modern microelectronics. Despite its size, Malaysia is moving forward in the Technology War.

Summary

The compendium of technology and competitors is now complete. The firms and states, and the national and international programs discussed in this and the last few chapters are shaping the battle and competing for victory in the Technology War. There are already fatalities such as SDS, Sitec (one of only two American manufacturers of silicon wafers), STC, Telex, Memorex, GE, RCA, Sperry Univac, AMD, GTE, and ITT, not to mention the wounded such as Trilogy, ZyMos, and so forth. More will come and it will be necessary for all of the competitors to track carefully the progress of all the others. As in any war, the surprises can come in any theater and at any time. Technical innovation can happen anywhere in the world. But even if we know the strategies of our opponents, what are the preferred options? In the next part we analyze some of the questions that face America in its preparation for the next series of battles in the Technology War.

PART IV

THE CHALLENGE

11

An American Strategy

One of the frightening things about the Technology War is that it is so complex to assess. Describing the problems confronting America is a very demanding task; evaluating the options available is a seemingly endless chore. In this chapter, we present a concise description of our view of the problems facing the United States and examine some of the more common alternative American views of these problems.

These alternative views of the future are disturbing since several of them conclude that no action is necessary, a conclusion with which we disagree. In response, we offer some recommendations in certain critical categories, the role of government, protectionism, national security, and education.

The American Problem

The United States is at an economic crossroads. The foundation of our traditional economic strength, the smokestack industries such as steel and automobiles, is crumbling. Those industries set the stage for a tremendous expansion of our economy and the foundation of wonderful technological innovation. They contributed to America's enormous advantage in the high technology markets after World War II. But the country's edge is gone. Our historic position as the preeminent manufacturing and exporting nation is under attack. Our manufacturing plant is aging. Competitors have absorbed our technology, exploited their own low labor cost structures and other domestic advantages, and have become threats to us on all fronts of the Technology War.

Are we content now to be an agricultural supplier? U.S. agriculture has become an efficient industry that employs only 3 percent of the population; it cannot deliver the jobs this country will need in the future. Should we depend upon a service-based economy? There were, in fact, some twenty million private sector service jobs created in the last few years. But to be in a service business is to be subordinate to the demands of others. These service jobs have other problems, too. They pay low salaries and may disappear altogether as additional manufacturing jobs are lost. And the service sector has never realized the growth in productivity that it expected from intensive computerization.[104]

The activities of our competitors are very serious. But we, ourselves, are our own worst enemy. Our lack of commitment, our failure to adopt a national policy, the continued emphasis on the support of old agricultural and manufacturing policies, the unconscionable and unrestrained growth in our trade and budget deficits, and our decrepit education system are terrible handicaps in the Technology War. The American system is presently geared to self-destruct.

We can no longer afford an uncoordinated economic and industrial approach to our myriad problems. Nor can we, as a nation, continue to ignore the development of technology because of the pressures of quarterly earnings reports. The current business preoccupation with speculation, mergers, golden parachutes, and poison pills is an unproductive use of both capital and human energy. The recent General Motors' payoff to H. Ross Perot of hundreds of millions of dollars is but one example of an outrageous use of crucial funds that could have been used to boost corporate competitiveness.[105] Our attention has been diverted from the important issues. We need only to consider unemployment, for example: The United States appears to have accepted a permanent level of unemployment two or more times that of Japan. We need to reject these figures of unemployment and work to reduce them.

On all fronts, if Americans cannot work more cheaply than others (such as the Chinese), nor more diligently than others (such as the Japanese or Koreans), then Americans must work more intelligently than others. But the United States' educational system is inferior to its competitors and is no more likely to produce workers with high technology skills in the future than it has in the past.

Technology may be, as President Reagan has said, the "parent of job creation."[106] But in order to realize these benefits, a good education is essential. With over 10 percent of the adult population functionally illiterate and unable to balance its checkbooks, it is clear that those Americans will not be programming the robots that will shoulder our future workloads. The predicted set of idle middle-class citizens that were supposed to enjoy the fruits of unlimited American economic expansion is unlikely to materialize. The problems confronting the United States are complex and pervasive.

Views of the Future

In our research and other activities, we have had the opportunity to discuss the topics in this book with policy makers, executives, economists, and scientists. Although there is agreement on many points, there are varying opinions on how to solve the problems facing the nation. We have collected these opinions and placed them in several positions, calling them the passive, the optimistic, and the pragmatic views.

The Passive View

In America today, people check their Japanese wristwatches for the time, drive to lunch in German cars, cash their checks at a foreign-owned bank, store the working files of their personal computers on a Fujitsu disk, and so forth. The computers which we used to prepare this book may have been manufactured in California, but the memory chips were Japanese and many of the American designed chips were manufactured abroad. Nevertheless, to all of this the passivist says, So what?

The passivists believe that America is a great and rich country with unlimited natural resources and a large, wealthy population. We manufacture plenty of important high technology products domestically, such as jet aircraft, satellites, and the TOW missile. Even the staggering national debt is nothing to lose sleep over. After all, we only owe it to ourselves. If necessary, the government merely needs to reignite inflation so that inflated dollars can be used to pay all the interest. What about the principal? Isn't that a long term problem? No, they say. In the long term, we will all be dead. Well, if not dead, perhaps some of the politicians mean that their elected terms will be over. The worst case, they say, is that we will just have a prosperous service-oriented economy.

The educational problems are dismissed as well. Privately, the passivist sympathizes with Japanese Prime Minister Nakasone, whose negative remarks about the ability to educate the American minorities received such extensive press coverage in the West. The passivist does not want to waste money on people he believes to be uneducable.

What about revitalizing our industrial plant? The passivists argue that the introduction of more technology into our economy will only worsen our unemployment problems. And after all, they say, many of the unemployed do not have the education, adaptability, or the motivation to learn.

World leadership? No problem, says the passivist. No one will push a nuclear power too far. And if our economy is strained because of defense, we can always bring the troops home from Korea, Japan, and Europe.

So, the passivist concludes, let's relax and enjoy our natural advantages. Perhaps it is time to acquire our realtors' licenses so that we can sell America's real assets to foreign investors.

The pragmatists think we should pay welfare taxes to maintain social stability, they think. Let's take advantage of our God-given standard of living and not embark on grandiose schemes to revitalize moribund industries.

Well, if this is a discouraging scenario, let us now hear from the positive thinkers.

The Optimistic View

The optimists concede that there has been a Technology War. But we are doing very well, thank you. We lead the world in Nobel Prizes and produce large numbers of innovative ideas. Besides, they say, the Japanese educational and science system is too confining, and as a result, the Japanese will never be able to compete with the West in innovation or basic research.

What if Japan was dumping chips in the United States? We have negotiated an agreement to make them stop. The agreement may have been too late to save the domestic memory chip industry, but as long as the government holds the Japanese to the agreement, another problem will be solved. Our auto industry is in trouble? Don't worry, they argue. Look at how Chrysler was salvaged. We can do anything we wish once we are aware of the problem. And, furthermore, we now have voluntary limitations on the number of cars that the Japanese can import into the United States. These limitations were a great victory for our American trade negotiators who convinced the Japanese to accept voluntary restrictions. We can be proud that they are not quotas, which would be un-American and smack of protectionism.

There's a loss of marketshare? That's no problem! We have the most advanced military and space technology in the world, optimists argue. Look at how we won the race to the moon. We will just spin off some more defense technology, build a supercollider, and create new jobs.

Nor is the deficit a cause for concern. Look at how our creative legislators solved the problem of the deficit by adopting the Gramm-Rudman bill, a model of legislation that guarantees fiscal reform.

Do we need coordinated government programs? Not at all, argues the optimist. We are so talented in our pluralistic approaches to problem solving that we can do it without the inefficient centralized economic management that the government would impose.

Although the reasoning is different for the optimists and the passivists, their agenda is the same: Do nothing. But we respectfully disagree! This entire conflict, and the issue of American competitiveness, is much too important to be left to chance or the mere sophistry of politicians.

A Pragmatic View

The pragmatists want to be specific about recommendations. They argue that the Technology War is very real. They are worried about comments like those of Professor Stephen Cohen, who said in a *U.S. News and World Report* :

> By 1995, there will be only high-technology industries in this country, whether they make parts, motors, insurance policies, or microchips.[107]

Pragmatists think of the technology race in terms of a marathon, that is, a test of endurance. The JTECH results and comparisons, in that context, are even more ominous, they say, because of the negative rate of change in America's comparative advantage in production. They fear the manufacturing expertise of the Japanese and consider the past battles with Japan over automobile imports, television sets, and chips as only early skirmishes in the Technology War. Pragmatists are convinced that the next round of confrontations will determine which great superpowers survive into the next century. They worry about America's chances.

The pragmatists believe that a poor educational system is a serious handicap in the Technology War and want to reform the elementary and secondary school systems. Salaries for teachers need to be raised, they argue, and far more competent teachers employed. They support the radical concept of rewarding teachers based on merit and are tired of endless arguments from mediocre teachers about job security and union rights. At the college level, they want the engineering schools to be revitalized. The pragmatists want the federal government, with help from the states and local industry, to re-equip the nations' laboratories. They condemn the business schools for preaching the short-term bottom line philosophy to MBAs and they want much more technology taught in the American graduate schools of business.

They are convinced that the growth of the service industries is the wrong way to solve America's problems. The pragmatists believe that only the restoration of American manufacturing operations will stop the decline, and they want a national commitment to manufacturing that could prove as expensive to the nation as the trip to the moon.[108]

Pragmatists argue that the drive to win can be instilled in people in many ways. They admire the Japanese practice of rewarding people who serve as technicians. The United States, they say, needs to glamorize manufacturing and the related processes. More people need to serve as manufacturing engineers; technicians need to be recognized for their important contributions in the production of goods.

The pragmatists are appalled at the bewildering condition of American science policy. Not too long ago, for example, President Nixon abolished the position of the Presidential Science Advisor, as well as the President's Scientific Advisory Committee. President Reagan appointed a Science Advisor but, according to members of his transition team, the White House sought appointees who would tend to avoid controversy and would maintain a low profile; the appointment was simply politically expedient. Pragmatists reject this philosophy and think America needs the strongest possible representation of science and technology in the White House!

They also are dismayed by the lack of other American coordination and cooperation in the Technology War. They wonder why there is no American MITI and reject the argument that the Defense Department does, or should, fill this role. Why, they ask, do the agencies of the U.S. government operate under an autonomous structure that was established at the time of the Industrial Revolution? It is time, they say, for the United States to stop preaching to the Japanese about the evil of their ways and to begin mimicking many of their practices.

The pragmatists want to confront the issue of defense in dealing with our opponents in the Technology War. They are tired of subsidizing foreign profits and want our allies to pay their fair share for their own defense. They worry that costly overseas garrisons for America's armed forces are addressing today's international political problems at the expense of tomorrow's problems. They know that reducing these costs will make more money available for American industrial development, while forcing our opponents to spend more of their money on defense and less on competing with us. Pragmatists also fear that the supply of technological personnel available to the domestic commercial sector is strained by the demands of defense. They worry that federal policy forces U.S. corporations to shift capital, which is in short supply because of the huge federal deficits, to defense production at the expense of commercial uses. They argue that the ultimate bastion of national security is a robust economy, not, for example, an excessive number of missiles that cannot be used.

Pragmatists sense that the real threat, still lurking behind our unchecked budget deficits, is runaway inflation and recession, if not depression. More importantly, they fear the loss of personal liberties that almost always have accompanied uncontrolled inflation in the past. They remember, for example, what happened to Germany after the mark collapsed in the 1920s.

Some pragmatic senators, such as Senator Jeff Bingaman (D., New Mexico), want to stop issuing defense contracts to overseas companies. They feel that the contracts impact negatively on the trade deficit and interfere with the development of our leading edge in research. Others, such as Senator Warren B. Rudman (R., New Hampshire), feel that the nation's military allies happen to be its fiercest competitors and that every dollar given them will come back to haunt us.[109] Senator John Glenn (D., Ohio) wants to establish a low limit on the entire overseas research and development awards made by the government; he cannot understand why America would want to subsidize foreign laboratories to compete with it.

One action considered by the pragmatists for dealing with Japanese competition is to invoke the methods of classical protectionism. For example, they suggest that Silicon Valley could compete with the Japanese if they had a "level playing field." They say that such market conditions can be achieved by imposing tariffs on Japanese imports, tariffs that must be maintained until America's marketshare in Japan is in balance with Japan's marketshare in the United States. However, they are ambivalent about such methods because they have an instinctive resistance to the evils of protectionism; they know that American steel producers, for example, have raised their prices every time the U.S. government invoked protectionist measures for their benefit.[110] The pragmatists want to protect the American consumer but feel America needs some more acceptable way out of its trade predicaments. They feel trapped between a rock and a hard place.[111]

Scientific freedom is another perplexing problem for the pragmatists. They know that the transfer of technology abroad is hurting the country and want to close the door on the "great technology giveaway." But they do not know how to do it without interfering with the traditional principles of scientific freedom and First Amendment rights. They think America has a difficult time ahead while it struggles with these issues, but they want to get the debate started now!

An American Strategy

The key to a prosperous economic future is to retain and expand our high technology edge. Our strategy must exploit those advantages of innovation and creativity. We must retrain our work force and improve our manufacturing sector. But, if we hope to finance the exploitation of high technology, the decline of American marketshares (the source of that financing) must cease and instead must start to increase. All of the interrelating factors, such as defense, trade, research, employment, and other societal issues, will turn on marketshare.

We also need to change our ideas about competition, employment, profits, and our political leadership. We need to elect leaders who are sensitive to these problems and whose political debts do not compromise their independence. Leaders who understand the relationships among intellectual property rights, industrial policy, trade policy, and tax and capital strategies are hard to find. The political system of America needs to offer such leaders and Americans need to elect them.

Clearly, a new strategy and tactics are required in the Technology War. In adopting these new tactics, America must discuss the tradeoffs between the cost of its defense, its social programs, the costs of education, and the organization of government entities. Our leaders must be sensitized to these critical issues, and a great American public awareness program is needed to educate the public and prepare it for some of the difficult choices ahead. Some of these choices strike us as apparent.

Government Policy and Organization

Most Americans would agree that the U.S. government must bear a major share of the blame for the decline of America's competitiveness. Americans disagree, however, on what should be done about competitiveness. Some believe the single greatest obstacle to an important government role in the Technology War is the government's inability to organize itself. They ask, why, for example, should we expect progress in this arena when there is precious little coordination today between the Departments of Defense, State, and Commerce and American industry?

They have a point. On first glance, reorganizing the government is not too attractive. After all, the government is not noted for its ability to accomplish much. It is huge and bloated, with a bureaucracy that uses archaic and ineffective policies and procedures. Most previous reorganizations have not been supported by the bureaucracy. Further, the government is preoccupied with the conflicts of the past and suffers an overdependence on the technologies and strategies of World War II. Its solutions to problems are usually far more costly than they would be if accomplished by private enterprise. Government has a reputation of abusing its power; there is every reason to assume that a nationalized computer industry, for example, would lead to a serious decay in the competitiveness of American computer companies.

On the other hand, there is no alternative to a strong government in terms of financial strength, the ability to afford risk, the power to command the attention of the people, and the ability to serve as the focal point for a national effort. No other entity can provide an umbrella under which companies can cooperate; no industrial organization can bring investment incentives, regulatory practices, and relief from old laws that are restraining America and offering loopholes to our competitors.

There are better ways to organize parts of our government. There is, in fact, one Japanese idea that is worth trying. Let's give serious consideration and support to the creation of a new Department of Commerce and Technology (DCT): that is, an American equivalent of MITI. Such a cabinet-level department could be composed of the Department of Commerce (and the National Bureau of Standards), the National Science Foundation, the Office of Science and Technology Policy of the White House, and the other government agencies and elements (other than DoD) concerned with research and development, industry, and trade.

DCT also would be similar to the British Department of Trade and Industry. It would integrate the charters of all of its component agencies and departments with the principal objective of winning the Technology War. The new department would help generate the needed national consensus, motivate the creation of more consortia, and stimulate more university and industry cooperation. It would represent all of our interests when the government establishes tax and capital policies. It would be charged with revitalizing the peoples' interest in education as a national asset, not just for vocational training or to make money. It would work with our beleaguered merchant semiconductor firms, who have now turned from technical to political solutions for their trade problems. It would have the ear of the President because it would be at the cabinet level.

Indeed, Congress has already received two proposals to establish just such a department. Senate bill S1233, introduced by Senator John Glenn, calls for the establishment of a Department of Industry and Technology; Representative George Brown (D., California) has proposed the establishment of a Department of Science and Technology.[112,113] Both these proposals have merit and deserve the most serious consideration. Congress can finance this reorganization if it is willing to revise its priorities to reflect the realities of the Technology War.

However distasteful, there are not many alternatives to a government reorganization. We need to focus our national efforts in the Technology War. We are reminded of the statement on competition made by Jerry Sanders, CEO of American Micro Devices: "Business is war and concentration of forces is an important strategy."[114]

Protectionism

The recent round of protectionism in the United States is aimed primarily at Japan because of the U.S.-Japan trade deficit. This negative trade balance is clearly related to the loss of American jobs. In fact, the U.S. trade deficit was approximately $60 billion in 1986 and the projected trade deficit for 1987 is estimated at about $100 billion, about $50 billion with Japan. Actually, the increase is not as dramatic as these figures suggest; the real change is only about 16 percent (the dollar figures are distorted by the recent fall of the dollar relative to the yen).

Nevertheless, it is still an enormous deficit and constitutes the mortgaging of the nation's future. As a result, the government is going to great efforts to combat the trade deficit. In one response, our trade partners, such as Japan and Korea, have been accused of various acts of unfair trade practices, especially dumping, which is the pricing of goods for sale at a price below cost. Although the International Trade Commission has ruled that Japan did indeed, dump semiconductor chips into the U.S. marketplace, the first question that we wish to address is whether a specific act of dumping is good or bad for the consumer.

One could take advantage of dumping. After all, the quality of (say) Japanese chips is high, and if the Japanese government and people wish to lose money on each sale, then we should accept their gift. The chips are a bargain, and the products that use them are less expensive and more competitive. On the other hand, dumping provides advantages for the Japanese because it increases their levels of employment and because it offers them economies of scale through larger production runs.

A more cogent point is that dumping costs American jobs. And, after the dumper has ravaged the competition and established an effective monopoly, prices presumably rise. But, in fact, this does not always happen. As we discussed in Chapter 5, in spite of an effective Japanese monopoly in the supply of television sets, prices have stayed low; with the increasing value of the yen, Korea has entered the market and undercut the Japanese.

What can the United States do about this complex problem? Obviously, it could impose quotas, import duties, domestic content laws, and so forth. But, these measures will produce only the usual consequences of protectionism, which are higher prices for domestic consumers, some jobs gained or retained (usually on a short-term basis), and, perhaps more threatening, unpredictable foreign reprisals. In protecting an industry like American microelectronics, for example, we must worry about the protection of inefficient manufacturers. The specter of an American ICL disturbs us greatly. Is it worth saving a troubled chip manufacturer if it makes American computer products noncompetitive? Protectionist acts will not help to solve America's long-term problems in competitiveness.

As this book goes to press, Congress is debating a 1987 trade bill. A critical aspect of the bill has become known as the Gephardt amendment which is named after its author, Representative Richard A. Gephardt (D., Missouri). This amendment could invoke sanctions against Japan or any other country with which we have a large trade deficit. It demands that such countries reduce their trade surpluses with the United States by at least 10 percent a year and it forces the U.S. government to negotiate the elimination of unfair trade restrictions on American suppliers. If a country is uncooperative, penalties are automatically imposed by the bill.

This kind of bill is undesirable because of its automatic response mechanism and simplistic ideas. It is better to have a flexible arsenal of weapons and tactics to deal with complex situations. Despite the attractiveness of simple solutions to complex problems, Washington needs to understand that there are no simple black and white solutions to the nation's fundamental trade and competitiveness problems.

Another simplistic method of protectionism proposed by some is to deny the use of our technology to our adversaries. This can be done in several ways. For example, we can prevent foreigners from attending high technology meetings, or foreign scientists and engineers can be excluded from U.S. universities and industrial research laboratories. But this would create hardships for America's engineering schools, where a high percentage of the students and faculty are from other countries. Also, Americans could expect the same treatment in response. Further, this would drive other countries to seek closer ties with Europe and the U.S.S.R., which would have significant negative foreign policy implications for the nation.

If America inhibits international technology transfer in such a manner, what happens to the free exchange of scientific information? These controls only serve to delay the dissemination of ideas and products; they cannot stop it. In fact, constraining technology transfer may have the undesired effect of impairing the development of American technology more than it impedes the flow outside the borders. Even today, America gains from the participation of foreign scientists. A disproportionately high share of scientific award winners in the United States are foreign. Further, we must ask if we would stop Americans from consulting abroad in support of such a policy. The services that we sell overseas are a substantial contributor to the credit side of the balance of payments ledger. It is unlikely the government is eager to abandon those credits.

Finally, it is prudent to note that our allies would surely not cooperate in restricting the flow of technology any more than they have before, such as in the Soviet pipeline or in the boycott of Libya. They will simply exploit the absence of Americans and steal the markets from the American suppliers and consultants.

From a constitutional point of view, these proposals are quite extreme. The consequences of the above actions are serious, and cannot possibly be anticipated. Although they would offer some relief to the trade problems of today, we recommend avoiding protectionism as much as possible. If the nation ever does choose to implement these techniques, they should be used only in a surgically precise and coordinated fashion and with the voluntary cooperation of American scientists and businessmen. The adverse effects of interfering with technology transfer might be greater than any advantage offered by protectionism. The Technology War will not be won by replacing it with a trade war.

The High Cost of National Security

One of the constraints restricting our funding of research and development for the commercial sector is the lack of money due to the enormous commitment of federal funds to the defense sector. Our national security is not only expensive in itself, it leaves few funds in the Treasury for commitment to our other national problems. We think it is time to re-examine our priorities. Now that the Technology War has become the new form of international strategic competition, the great emphasis on military defense spending needs review. This need for review is amplified by the demands of the Star Wars program.

Besides its apparent contribution to defense, SDI is frequently defended on the grounds that it will produce technology of value to the commercial sector. To be more specific, the argument is offered that, as a side-effect of developing the military technology there will be a great commercial payoff from SDI. This will justify the expense and SDI will therefore prove cost effective from a domestic point of view. But, such a spinoff argument is used by the advocates of any large scientific program.

The argument also requires an act of faith. In fact, the best way to obtain the civilian scientific advances that are needed for commercial competitiveness is to work directly on the industrial problems. Indeed, the spinoff argument may no longer have any validity, since a search for examples of successful spinoffs from recent defense and space research programs yields few returns. The Boeing 707 is the prototypical example, but there are few examples today of such commercial payoffs from defense research. The potential spinoff has been reduced as the Reagan administration has increased the government's concentration on defense and imposed greater secrecy on all of its activities. The jury is still out on the spinoff argument for SDI.

One case against massive defense research and development, in general, and Star Wars, in particular, is that it diverts our scarce technical talent from the commercial sector to the defense sector. Defense, for example, already employs 40 percent of America's scientific talent.[115] And, while earlier military and space programs have included programs for training more engineers and scientists, SDI carries no such burden. SDI may be too expensive in these terms, given the expected commercial benefits.

In addition, the issue of big science versus little science must be faced in any discussion of SDI. Examples of other big science projects, sometimes called megaprojects, are NASA's space program and the superconducting supercollider that will be used by particle physicists. This collection of projects could cost well over $25 billion. Combined with SDI, the diversion of potential research funds to big science projects could be $50 billion. But, as experience shows, many of the innovative new ideas, especially in the new emerging technologies, come from small projects. It is not clear we can afford the cost to our smaller efforts of all of these large projects.

More spending is needed, to re-equip the university laboratories. Although laboratory renovation may be a $10 billion investment, it must be made or our engineering graduates will continue to be ill prepared for modern industry.[116] If this modernization cannot be supported because of the high cost of defense, we may have to find a system whereby industry assumes the burden for educating engineers.

National security constraints on American exports have to be relaxed as well. The controls have not proven effective, for example, and the recent National Academy of Science (NAS) study on export controls found that, in most cases, the costs to American suppliers in terms of jobs, profits and marketshare of export controls far outweighed any of the benefits.[35] In fact, more than half of the American firms surveyed by the NAS panel in electronics, aircraft, and machine tools (representing over 15 percent of the total U.S. high technology sales) reported a serious loss of sales as a consequence of export controls and that they expected such problems to increase over the next two years. The NAS report calling for wide-ranging abandonment of controls must be adopted. We cannot afford to summarily accept former assistant secretary of defense Richard Perle's assessment of this report that the studies are "sheer rubbish."[117]

To be competitive in the Technology War, the U.S. needs to spend some 2.7 percent of the GNP on commercially related research and development; this is comparable to what the Japanese spend. Although we are spending vast sums in related areas, not enough of our current national spending on research is focused properly. An excessive amount is committed to defense problems. We simply cannot depend on spinoffs from the U.S. defense research to the private sector and we desperately need a higher level of capitalization per researcher. The nation must choose between investment in defense and investment in competitiveness.

The issues are complex and demanding. The choices are hard and cannot be made by any group other than the Congress of the United States. We cannot expect the Department of Defense, despite its relatively superior technological competence, to be objective in this matter. Congress needs to determine its priorities, and then, to meet those priorities in a program that balances the interests of national security and the commercial and scientific communities with the needs of the Technology War.

Education

The case for educational reform and federal support is so strong and pervasive that the arguments in favor need not be repeated here. On the other hand, there are good arguments against massive federal investment in education: We have had similar investments in the past and they have demonstrated that simply throwing money at the problem does not work. Our teachers are no better than they were twenty years ago, our children receive a poorer education than we did, and the teachers' unions are as vociferous as ever about their right to defend the pension rights and working conditions of the teachers of America.

School districts could strive for excellence in their educational systems on a local basis. But, under these conditions, national uniform standards would not be applied to students or to teachers. Further, many school districts in America do not have enlightened voters, some districts simply lack the resources, and so, to us, there is no alternative to the federal support of education. We find it sad that House Speaker James Wright, in his response to President Reagan's 1987 State of the Union Address, felt it necessary to call attention to the shortfalls in the President's proposed education budget. Wright was especially astute when he said, "However splendid our weapons, we won't be first in defense or trade if we settle for second best in education."[118]

We think that America has to put performance standards in place and then enforce them. The Japanese accept such demands. So can we. There must be federally standardized examinations to ensure a minimum level of competence of both high school graduates and teachers. America must stop the practice of graduating students whom it has not educated.

At the college level, our engineering schools must be salvaged. New incentives are essential that will attract, and retain, top quality U.S. faculty members as well as top rank U.S. graduate students into full time graduate study. The educational reforms used in the technical schools during the Sputnik era may need to be used again.

Conclusion

We do not presume to have all the answers to the perplexing questions and tradeoffs that we have discussed. In a complex situation, the "do nothing" approach of the optimists or the passivists may seem attractive. We think that is a dangerous perspective. The problems demand action, but action demands consensus.

We fear that the prospects for a national consensus on the tradeoffs among scientific freedom, technology transfer, protectionism, and competitiveness are poor. Reaching a consensus is no simple task in a political system which is dominated by defense issues and in which government and industry have a poor dialogue. Such consensus can only be orchestrated by a respected and informed leader in the White House. In the last few years of the beleaguered Reagan administration, there is not much hope for such leadership.

With prospects of a weak government, we can only hope that the Technology War becomes a key issue in the 1988 campaign. But even with enlightened politicians, the country faces a difficult road. The attitudes of the American people need to be changed, and changing people's behavior is the most difficult task of all.

People will respond to incentives, but only slowly, and governments, no matter how enlightened the leaders, are even slower to change the incentives or, alas, to abolish the disincentives.

We do know that the pace of change can be quickened. This is the century of change. Great social, technological, and political progress has been demonstrated and accomplished. Consider how Russia was transformed by the Russian Revolution of 1917; how America was motivated, for example, by Pearl Harbor and Watergate. All it takes is a national will and purpose. Unfortunately, it takes more than words. Robert E. Kahn, a former director of a DARPA office, who has contributed as much as anyone to the development of advanced computer technology in the United States has said

> The number of people ... who have thought long and hard about this problem in terms of national strategy is very small. Even if we could *will* the technology into existence today, would we find the national purpose to succeed in exploiting it competitively?[119]

The war will be won when we have found that national purpose, when *our* unemployment rates are the envy of the world, when we have restored the trade surpluses at the expense of our trading partners, and when the dollar is not a prisoner of war in trade battles. The war will *remain* won only as long as our population is technologically competitive. Coordination, cooperation, education, and hard work are the long-term keys to the leadership of the twenty-first century. It begins there and it ends there.

Bibliography

1. *Information Processing in the United States.* AFIPS Press, Reston, VA, July 1985. Revised Edition.

2. Franklin M. Fisher, John J. McGowen, and Joel E. Greenwood. *Folded, Spindled and Mutilated, Economic Analysis and U. S. vs IBM.* MIT Press, Cambridge, MA, 1983.

3. Walter B. Wriston. "Gnomes, Words and Policies." *Harpers,* September 1985. A speech given to the Executive's Club of Chicago on May 8, 1985.

4. Arthur C. Clarke. *Ascent to Orbit,* chapter 9. John Wiley & Sons, New York, 1984.

5. Luigi Barzini. *The Europeans.* Simon & Schuster, New York, 1983.

6. K. Davis and R. L. Blomstrom. *Business, Society, and Environment: Social Power and Social Response.* McGraw-Hill, New York, 1971.

7. Edwin O. Reischauer. *The Japanese.* Belknap Press of Harvard University Press, Cambridge, MA, 1981.

8. Mark Crawford. "Japan's U.S. R&D Role Widens, Begs Attention." *Science,* 233(4761):270–272, July 1986.

9. Stephen T. McClellan. *The Coming Computer Industry Shakeout:Winners, Losers and Survivors.* John Wiley & Sons, New York, 1984.

10. Ezra Solomon. "The Anxious Economy." In *The Portable Stanford*, Stanford Alumni Association, Stanford, CA, 1975.

11. Sun Tzu. *The Art of War.* Delacorte Press, New York, 1983. Edited and with a Forward by James Clavell. The original was published in the 6th century B.C.

12. Joe Rudzinski. "Samurai Strategy in Modern Japan." *SRI Insights*, 1(1):5, Spring 1986.

13. Kunio Murakami. "Archery Discipline and Fifth Generation Computer Research." *ICOT Journal*, (2):39–41, September 1983.

14. D. Brandin, J. Bentley, T. Gannon, M. A. Harrison, J. Riganati, F. Ris, and N. Sondheimer. *JTECH Panel Report on Computer Science in Japan.* Technical Report, Science Applications International Corporation, 1200 Prospect Street, La Jolla, CA 92037, 1984.

15. Toshio Yukuta, Eiji Kitihara, Warren Hegg, Ivor Brodie, Julius Murray, and Terrence Cullinan. "Science Discovery in Japan: Dawn of a New Era." *Science85*, 6, 1985. Special advertising section.

16. J. Nishizawa. Japanese Patent No. 39-6404. Japanese Patent Office, 1964.

17. Steve Lohr. "The Japanese Challenge." *The New York Times Magazine*, July 1984.

18. *AFIPS Washington Report.* AFIPS Quarterly Newsletter, April 1986. Vol. XII, No. 4.

19. Eric T. Bell. *Men of Mathematics.* Simon & Schuster, New York, 1937.

20. Edward Feigenbaum and Pamela McCorduck. *The Fifth Generation.* Addison-Wesley, Reading, MA, 1983.

21. Michael Borrus, James Millstein, and John Zysman. *Responses to the Japanese Challenge in High Technology: Innovation, Maturity and US-Japanese Competition in Microelectronics.* Technical Report, Berkeley Roundtable on the International Economy, University of California, Berkeley, July 1983.

22. J. Morris. "France Plans DP Sell-Off." *Datamation*, 32(10):64–68, May 1986.

23. A. R. Chandler. *The Clash of Political Ideas.* Appleton-Century-Crofts, New York, 1949.

24. Office of Technology Assessment. *International Competitiveness in Electronics.* Technical Report, U.S. Government, U.S. Government Printing Office, Washington, DC, November 1983. (OTA–ISC–200).

25. Herbert L. Dreyfus and Stuart E. Dreyfus. "Competent Systems: The Only Future for Inference–Making Computers." *Future Generation Computer Systems*, 2:223–243, 1986.

26. J. Nevins, J. Albus, T. Binford, M. Brady, N. Caplan, M. Kutcher, P. J. MacVicar-Whelan, G. L. Miller, L. Rossol, and K. Schutz. *JTECH Panel Report on Mechatronics in Japan.* Technical Report, Science Applications International Corporation, 1200 Prospect Street, La Jolla, CA 92037, 1985.

27. David Kahn. *Hitler's Spies.* Macmillan, New York, 1978.

28. W. J. Spencer. "Technology Transfer in the U.S. – Challenges and Opportunities." In *U.S. China Conference on Technology Transfer*, American Council of Learned Societies, National Research Council, Social Science Research Council, National Academy Press, Washington, DC, November 1985.

29. *The Findings of the Public Cryptography Study Group.* American Council on Education, Washington, DC, May 1981.

30. G. Gervaise Davis III. *Software Protection.* Van Nostrand-Reinhold, Princeton, NJ, 1985.

31. *Moving Singapore into the Information Age.* National Computer Board Yearbook, Singapore, 1984-1985.

32. *Proposal with Respect to Rearrangement of the Foundation for Software–Aiming at Securing Legal Protection of Software.* Industrial Structure Council, Information Industry Committee, Tokyo, Japan, December 1983. Interim Report.

33. *A Report on Japanese Legal Protection of Software.* The Software Legal Protection Committee, Japan Software Industry Association, Tokyo, Japan, May 1983.

34. Bruce Gilchrist and Milton R. Wessel. *Government Regulation of the Computer Industry.* AFIPS Press, Reston, VA, 1972.

35. National Research Council. *Balancing the National Interest.* National Academy Press, Washington, DC, 1987.

36. Harry Rositzke. *The KGB: The Eyes of Russia.* Doubleday, New York, 1981.

37. Michael Wels Hirschon. "Ohio Firm is Entangled in Confusion over U.S. Policy on High Tech Exports." *Wall Street Journal,* CXVI(28):16, February 10, 1987.

38. David Bergamini. *Japan's Imperial Conspiracy.* William Morrow & Co., New York, 1971.

39. Chalmers Johnson. *MITI and the Japanese Miracle.* Stanford University Press, Stanford, CA, 1982.

40. Lee Iacocca and William Novak. *Iacocca: An Autobiography.* Bantam Books, New York, 1984.

41. William F. Baxter. *Antitrust and Technological Change: A Problem of Synthesis.* Speech before The Town Hall of California, October 7, 1983.

42. William F. Baxter. *Antitrust Law and the Stimulation of Technological Invention and Innovation.* Conference on Government Organization and Operation and the Role of Government in the Economy, July 1953. University of San Diego, San Diego, CA.

43. Michael Borrus, James Millstein, and John Zysman. *US-Japanese Competition in the Semiconductor Industry*. Policy Paper 17, Institute of International Studies, University of California, Berkeley, 1982.

44. Janice C. Simpson. "On the Road to the 21st Century." *Time Magazine*, 128:35, July 7, 1986.

45. *How to Get Americans to Sock Away More*. Business Week, April 13, 1987.

46. Murray L. Weidenbaum. *Business, Government and the Public*. Prentice Hall, Englewood Cliffs, NJ, second edition, 1981.

47. Barbara W. Tuchman. *Stilwell and the American Experience in China*. Macmillan, New York, 1970.

48. Winston Churchill. *Winston Churchill: British Bulldog*, chapter XLI. Exposition Press, London, February 1955.

49. K. Kato. *The White Paper: Defense of Japan*. The Japan Times Ltd., Tokyo, 1985.

50. National Science Board. *Science Indicators, the 1985 Report*. Technical Report 038-000-00563-4, U.S. Government, Superintendent of Documents, U.S. Government Printing Office, Washington, DC 20402, 1985.

51. National Commission on Excellence in Education. *A Nation at Risk: The Imperative for Educational Reform*. Technical Report, U.S. Government, Superintendent of Documents, U.S. Government Printing Office, Washington, DC 20402, April 26 1983.

52. Lawrence P. Grayson. "Education and America's Industrial Future." *Computer*, 19(6):10–17, June 1986.

53. Bruce L. Smith, editor. *The State of Graduate Education*. Brookings Dialogues on Public Policy, Brookings Institution, Washington, DC, 1985.

54. Erich Bloch. "Basic Research and Economic Health: The Coming Challenge." *Science*, 232(4750):595–599, May 1986.

55. Robert G. Snyder. "Some Indicators of the Condition of Graduate Education in the Sciences." In Bruce L. Smith, editor, *The State of Graduate Education*, pages 31–56, Brookings Institution, Washington, DC, 1985.

56. Paul Doigan. "Engineering Education, Fall 1982." *Engineering Education*, 74(1):18–20, October 1983.

57. Harvey Brooks. "The Outlook for Graduate Science and Engineering." In Bruce L. Smith, editor, *The State of Graduate Education*, pages 181–188, Brookings Institution, Washington, DC, 1985.

58. F. Karl Willenbrock. "The Status of Engineering Education in the United States." In Bruce L. Smith, editor, *The State of Graduate Education*, pages 85–96, Brookings Institution, Washington, DC, 1985.

59. Deborah Shapley and Rustum Roy. *Lost at the Frontier*. ISI Press, Philadelphia, PA, 1985.

60. D. Eleanor Westney, K. Sakakibara, and Dan P. Trullinger. *Comparative Study of the Training, Careers, and Organization of Engineers in the Computer Industry in Japan and the United States*. M.I.T.–Japan Science and Technology Program, October 1984.

61. *Curriculum '68: Recommendations for Academic Programs in Computer Science*. Association for Computing Machinery, New York, 1968.

62. *Curriculum '78: Recommendations for the Undergraduate Program in Computer Science*. Association for Computing Machinery, New York, 1978.

63. Department of Industry. *A Programme for Advanced Information Technology*. Technical Report, Alvey Directorate, British Department of Industry, London, 1982.

64. Judith Miller. "Frenchman Assailed for Denying Nazi Crimes." *New York Times*, CXXXV(46,804):11, June 13, 1986.

65. Ezra Vogel. "The Changing Nature of Information Societies." *ICOT Journal*, (7):2–11, March 1985.

66. Thomas H. Etzold. *Defense or Delusion? America's Military in the 1980's.* Harper & Row, New York, 1982.

67. U.S. Bureau of the Census. *Statistical Abstract of the United States 1987.* Technical Report, U.S. Government, Superintendent of Documents, U.S. Government Printing Office, Washington, DC 20402, 1987. (107th edition).

68. Defense Advanced Research Projects Agency. *Strategic Computing, New-Generation Computing Technology: A Strategic Plan for its Development and Application to Critical Problems in Defense.* Technical Report, U.S. Department of Defense, Washington, DC, October 28 1983.

69. M. Mitchell Waldrop. "Resolving the Star Wars Software Dilemma." *Science*, 232(4751):710–713, May 1986.

70. R. Smith. "Hicks Attacks SDI Critics." *Science*, 232(4749):444, April 1986.

71. J. C. van Vliet. "STARS and Stripes." *Future Generation Computer Systems*, 1(6):411–416, December 1985.

72. Edward Lieblein. "The Department of Defense Software Initiatives – A Status Report." *Communications of the ACM*, 29(8):734 – 744, August 1986.

73. *Report of the Federal Coordinating Council on Science, Engineering, and Technology Panel on Advanced Computer Research in the Federal Government.* June 1985.

74. Michael A. Harrison. "The Role of Government, Industry, and Universities in Support of Research and Development in the Field of Knowledge Systems." In Bernard A. Galler, editor, *Cooperation through Competition*, U.S.–Japan Seminar, Tsukuba, Japan, June 1985.

75. *IBM 1985 Annual Report.* International Business Machine Co., Armonk, NY, 1985.

76. The Gartner Group, Inc. *Mainframe Market Share: IBM Continues to Gain.* Newsletter, Stamford, CT, May 1985. File: M-550-341.

77. The Gartner Group, Inc. *Software Management Strategies, IBM's Software Strategies: Competitive Implications.* Newsletter, Stamford, CT, May 1985. (File: S-007-055.1).

78. Rex Malik. *And Tomorrow the World? Inside IBM.* Millington Ltd., London, 1975.

79. Arthur L. Robinson. "A Superconductivity Happening." *Science,* 235(4796):1571, March 27, 1987.

80. M. D. Fagen, editor. *A History of Engineering and Science in the Bell System: The Early Years (1875-1925).* AT&T Bell Laboratories, Murray Hill, NJ, 1975.

81. Federal Communication Commission. "FCC docket number 16979, final decision and order." In Bruce Gilchrist and Milton R. Wessel, editors, *Government Regulation of the Computer Industry,* pages 121–153, AFIPS Press, Reston, VA, 1972. September 24, 1971.

82. F. M. Smits, editor. *A History of Engineering and Science in the Bell System: Electronics Technology (1925-1975).* AT&T Bell Laboratories, Murray Hill, NJ, 1985.

83. Dennis M. Ritchie. "Reflections on Software Research." *Communications of the ACM,* 27(8):758–760, August 1984.

84. Tim Beardsley. "From Monopoly to Competition." *Nature,* 319:90–91, January 9, 1985. Vol. 319, pp 90–91.

85. T. Kurihara. "Software: Critical Decisions." In *Proceedings of the Second Software Engineering Standards Applications Workshop,* 1983.

86. Philip Hunter. "In Europe, the "Bunch" Hope to Connect to Top Banana IBM." *InformationWEEK,* 30, May 5, 1986.

87. Tom Ewing. "The BUNCH Report." *Information WEEK*, (057):29–53, March 17 1986.

88. Katherine D. Fishman. *The Computer Establishment.* Harper & Row, New York, 1981.

89. Regis McKenna. *The Regis Touch.* Addison-Wesley, Reading, MA, 1985.

90. Gary Jacobson and John Hillkirk. *Xerox, American Samurai.* Macmillan, New York, 1986.

91. J. P. Donlon. "Technology Venturer Bobby Inman Nears the First Hurdle." *Chief Executive*, (35):30–32, Spring 1986.

92. William Ouichi. *The M-Form Society.* Addison-Wesley, Reading, MA, 1984.

93. H. Wieder, W. Spicer, R. Bauer, F. Capasso, D. Collins, K. Hess, H. Kroger, R. Scace, W. T. Tsang, and J. Woodall. *JTECH Panel Report on Opto-& Microelectronics.* Technical Report, Science Applications International Corporation, 1200 Prospect Street, La Jolla, CA 92037, 1985.

94. Makoto Nagao, Toyoaki Nishida, and Junichi Tsujii. "Dealing with Incompleteness of Linguistic Knowledge in Language Translation: Transfer and Generation Stage of Mu Machine Translation Project." In *Proceedings of the Conference*, Association for Computational Linguistics, July 1984.

95. Pamela Archbold and Parker Hodges. "The Datamation 100." *Datamation*, 32(12), June 1986.

96. Denji Tajima and Tomoo Matsubara. "Inside the Japanese Software Factory." *Computer*, 17(3), March 1984.

97. *Towards a European Strategic Program for Research and Development in Information Technologies.* Communication from the Commission to the Council of the Commissioner of the European Communities, May 1982. Brussels, Belgium.

98. Anonymous. Private Communication. August 1986.

99. Gao Lulin. "Transforming China's Traditional Industries." In *U.S. China Conference on Technology Transfer*, American Council of Learned Societies, National Research Council, Social Science Research Council, National Academy Press, Washington, DC, November 1985.

100. Xu Zhaoxiang. "The Status of S & T Policy Research in China." In *U.S. – China Conference on Technology Transfer*, American Council of Learned Societies, National Research Council, Social Science Research Council, National Academy Press, Washington, DC, November 1985.

101. Teng Teng. "The Reform of the R & D Management System and Development Strategies in China." In *U.S.– China Conference on Technology Transfer*, American Council of Learned Societies, National Research Council, Social Science Research Council, National Academy Press, Washington, DC, November 1985.

102. Louis Kraar. "Reheating Asia's "Little Dragons"." *Fortune*, 113, May 1986. pp. 134–140.

103. *The Singapore Economy: New Directions*. Report of the Economic Committee, Ministry of Trade and Industry, Republic of Singapore, February 1986.

104. Norman Jonas. "Can America Compete?." *Business Week*, 46, April 20 1987.

105. *A Heckled Chairman Defends His Program*. New York Times, May 23, 1987.

106. Gerald M. Boyd. "Reagan Hails Technology as Key to U.S. Development." *New York Times*, CXXXVI(47,105):8, April 10, 1987.

107. Lucia Sollorzano. "Jobs of the Future." *U.S. News and World Report*, 99(26):40–48, December 23 1985.

108. Otis Port and John W. Wilson. "Making Brawn Work with Brains." *Business Week*, 56, April 20 1987.

109. S. Diamond. "U.S. Control of Arms Sales Weak and Ineffective, Senate Unit Told." *The New York Times*, CXXXVI(47,057):2, February 21, 1987.

110. Warren T. Brookes. "Protectionism's Failure in Steel Industry." *San Francisco Chronicle*, (52):A 6, March 18, 1987.

111. Martin Tolchin. "Foreign Investment in U.S. Mutes Trade Dispute." *The New York Times*, CXXXVI(47,044):30, February 8, 1987.

112. John Glenn. *S1233: Department of Industry and Technology.* Congressional Record, United States Congress, Washington, DC, May 19, 1987. Pages S6753-S6764.

113. *Urges Science Dept. for Cabinet.* Electronic News, Washington, DC, May 11, 1987. Page 73.

114. Don Clark. "AMD, Monolithic in $422 Million Deal." *San Francisco Chronicle*, (90):37, May 1, 1987.

115. Lester C. Thurow. "The Economic Case Against Star Wars." *Technology Review*, 89(2):11, 15, Feb.–Mar. 1986.

116. Steven Solomon. "Will Old Machines Kill New Ideas?." *New York Times*, CXXXVI(47,163):F 4, May 1987.

117. New York Times News Service. "Top Pentagon Official Rejects Tech Report." *Sunnyvale Times Tribune*, C 10, February 5, 1987.

118. Linda Greenhouse. "Reagan's Address Provokes a Highly Partisan Reaction." *New York Times*, CXXXVI(47,033):A 17, January 28, 1987.

119. David H. Brandin. "The Challenge of the Fifth Generation." *Communications of the ACM*, 25(8):509–510, August 1982.

Index